BOSS WHISTLE

THE COAL MINERS OF VANCOUVER ISLAND REMEMBER

BOSS WHISTLE

THE COAL MINERS OF VANCOUVER ISLAND REMEMBER

Revised Edition

by

Lynne Bowen

Nanaimo and District Museum Society
And
Rocky Point Books
Nanaimo, B.C.
2002

ISBN 0-9697407-1-9

National Library of Canada Cataloguing in Publication data

Bowen, Lynne, 1940-
Boss whistle

Co-published by: Nanaimo and District Museum Society.
Includes bibliographical references and index.
ISBN 0-9697407-1-9

1. Coal mines and mining—British Columbia—Vancouver Island—History. 2. Coal-miners—British Columbia—Vancouver Island—History. 3. Strikes and lockouts—Coal mining—British Columbia—Vancouver Island—History. I. Nanaimo and District Museum Society. II. Title.
HD9554.C23B75 2002 338.2'724'097112 C2001-911726-4

Editor: Rhonda Bailey
Layout: Phantom Press
Cover design: Andrew Bowen
Maps and Photographic Restoration: Andrew Bowen

Published by
ROCKY POINT BOOKS
4982 Fillinger Crescent, Nanaimo, B.C., V9V 1J1
And
NANAIMO AND DISTRICT MUSEUM SOCIETY
100 Cameron Road, Nanaimo, B.C., V9R 2X1
Tel: (250) 753-1821; Fax: (250) 518-0125;
E-mail:ndmuseum@nanaimo.museum.bc.ca

Distributed by
SANDHILL BOOK MARKETING LTD.
#99-1270 Ellis Street, Kelowna, B.C., V1Y 1Z4
Tel: (250) 763-1406; Fax: (250) 763-4051; E-mail:sandhill@direct.ca

Printed by
FRIESENS

Cover Photo: Kathleen Berry, donor, Nanaimo District Museum 03-4

To the coal miners of Vancouver Island,

their loved ones, and their descendants.

Contents

Preface

Twenty-one years ago I began to work with the Coal Tyee Society, a group of coal miners and their friends who entrusted me with their stories after Myrtle Bergren, the woman who had inspired them, died in a car accident. The Society disbanded in 1990, but the stories live on.

There have been ten printings of the first edition of *Boss Whistle*. It resides in the collections of major libraries across Canada. Well-thumbed copies occupy the shelves of museums and archives even as new information and photographs continue to materialize.

I have enjoyed hearing from many people, some to say thank you for the book, some to offer the use of documents and photographs and a few to tell me I made mistakes. I am grateful to them all. The revised edition is for the people who read the original and want to know more and for the people who will read the coal miners' story for the first time.

The revised edition of *Boss Whistle* owes much to the people whose names are in the preface of the original, but in addition I want to thank Hazel Galloway, Marcia Galloway, Gladys Campbell, Annie Dempster Moore Clark, Kathy and Garth Gilroy, Graham Holland, Thora Howell, Joan Auchinvole Hurn, Shirley Johnson, Dorothy Johnstone, Gordon B. McLean, Olive Parker, Frank and Joanne Ueberschar, B. Parker Williams, Trevor Shields, and the late David Ricardo Williams.

Rocky Point Books and the Nanaimo and District Museum Society are the co-publishers of this new edition. I thank the museum's board of directors for its moral and financial support and John Manning for preparation of the contract. Debbie Trueman, the museum's general manager, suggested the partnership and gave good counsel. Richard Slingerland produced an important photograph. Barb Lemky, curator of the Cumberland Museum, and Jack Bartell were generous with their time and expertise. Christine Meutzner and the staff of the Nanaimo Community Archives clarified many details. With all this expert help, any mistakes of fact or interpretation that remain are mine alone.

In the past twenty years a generation has grown up knowing only the metric system of measurements. Since the miners used only imperial and young readers know nothing but metric, I have decided to use metric for the text and imperial in the miners' quotations. Perhaps in another twenty years this dual measurement system will no longer present a problem.

I have retained such terms as "Chinamen," "Nigger," and "Japtown" when the speakers use them. As offensive as these terms sound to our ears, they were common during the time when coal mining was active on Vancouver Island. I believe it is important for readers of history to understand the careless but often naïve use of these words in context.

I am fortunate that Rhonda Bailey, a long-time friend of *Boss Whistle*, consented to edit the manuscript, and that page design and layout expertise were available close to home in the person of Linda Hildebrand at Phantom Press. Even closer to home, my son, Andrew Bowen, drew the maps, designed the cover and rehabilitated some of the archival photographs. Barbara Lyall, Ken Lyall, John Manning, Marjorie O'Callaghan, John O'Callaghan, Andrew Bowen, and Debbie Trueman proofread the galleys. Gerhard Aichelberger shepherded the production of the finished book.

Readers wishing to learn more about Island coal mining can refer to my books *Three Dollar Dreams* (Oolichan Books, 1987) and *Robert Dunsmuir, Laird of the Mines* (XYZ Publishing, 1999). It is an honour to be associated with the coal miners' story.

Lynne Bowen
Nanaimo, B.C.
December 2001

Maps

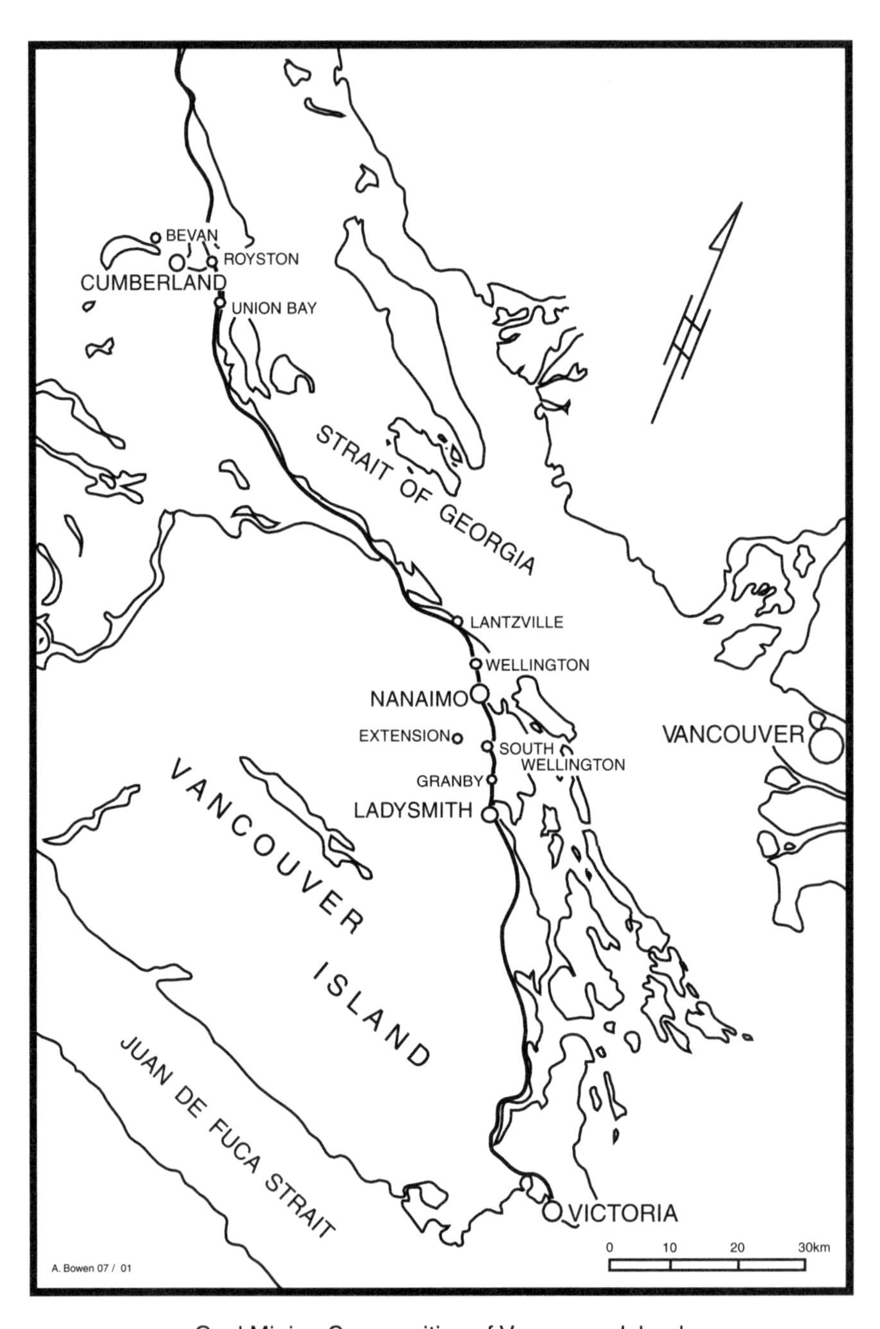

Coal Mining Communities of Vancouver Island

CHAPTER I

Boss Whistle

The cage plummets downward so fast that a miner closing his eyes feels as if he is floating. It shudders to a halt at shaft bottom. Steel doors crash open; the cager urges the men off; their places are taken by the shift just ended. Rush. Rush. Get the man trips over so the coal can go again. Keep the coal moving.

The brightly lit shaft bottom, hundreds of feet beneath the surface, bustles with activity. Coal laden cars clatter over track, metal wheels on metal rails; mules bray as they resist harnessing, the hitching chains rattle and clash; bosses shout orders, marshalling crews, chastising novices: "Move on, don't waste time."

For the crews disappearing into the labyrinth of the mine, some on empty cars behind small electric locomotives, some on foot, the clamour dies away. At the face there will be the noise of machinery and the deep *whump* of shots detonating, but as each miner heads for his assigned place, there are a few moments for reflection.

As the logger knows the woods and the fisherman knows the sea, so the miner knows the world underground intimately. He knows the coal in all its variations and the rock supporting it and embedded between its seams. He knows a solid roof from a friable one and why both can be dangerous. He quickly acquires a sense of direction and distance in the maze of tunnels. The sounds a mine makes are familiar, and he can detect the subtle difference between what is merely the roof settling or the timbers creaking and what is unusual and warns of danger.

He knows the mine will kill him if he lets down his guard, but like a fighter who respects his opponent, he keeps a wary eye out for the lethal punch and can sometimes counter it. The mine can kill with a fall of rock or coal, with a flood, with a runaway car, with an explosion of gas. The miner knows that the gas bleeds out of the coal and can explode at the slightest spark or flame. The ventilation system of the mine, designed to drive away the gases, is the miner's best defence, and he is more secure as he feels the air, sucked by a massive fan, sweep by his face.

At the working face the only lights are those carried by the miners. In the vast and deep darkness the lights are feeble. Despite the sweep of the ventilation system, the air is choked with fumes from the blasting. Men cough and curse. They drill and dig and heave heavy timbers. They work on their feet and their knees and their bellies. Digger and back-hand, chunker and mucker, driver and rope rider, black-faced, sweating, joking, and cursing.

For each of the thousands of men who worked in the coal mines of Vancouver Island, the experience was different. The romantics found the mine hospitable, its darkness and even temperature reassuring, the black coal beautiful; the pragmatic accepted the lack of sunlight, the wetness, the danger as necessary in the pursuit of a living; the sensible saw the realities of the mine and left, never to return.

> I first went down as a winch boy and I was with another young fella. The first day in the mine, the first trip I pulled up from the face was a dead mule that had been killed on the shift before. This other young fella was workin' farther up and I would pull one trip up and he picked it up and took it up to the next winch. The second trip I pulled up, there was a man sitting in the box of the car and he had been hit in the head with a fall of coal. He was quite a mess with black face and blood running down and the other young fella when he pulled that man up, he left the mine. Quit right there. He just couldn't take it.

For most men, it was a matter of adjustment. The young winch boy or rope rider overcame his fears and soon became a seasoned underground worker. The process of adjustment required an iron nerve, a determination never to give in to fear, and then the development of an ability to use the mine to his advantage.

> One place I heard this rumble away far away and I thought, "Oh Christ, the mine's exploding." It's kind of eerie. I know lots of times you're working on a winch and you're all by yourself and you're sitting there waiting for your rope rider to give you the signal to do something and you hear little bits of coal tumbling, maybe falling off the wall or something. "Jesus, is the mine caving in?" Or maybe you hear a rat scurrying around or you hear something move . . .

> You couldn't see. You had to go by sound. You watched your rope too. You knew if you was comin' near the bottom or near the top by the coils on your rope. You could hear the cars comin'; after the loads would pass the empties, the sound would be more clear. And then you would stand up and hold on to the brake and watch the trip comin' over the knuckle.

The lack of light accentuated the sounds – the useful and the dangerous – but most miners were never bothered by the darkness even when their headlamps went out, which was often. A man could call out to someone nearby or he could follow a rail with his foot until he came to where there were other miners.

> I never got lost underground. My light would go out all the time, but if you didn't have a light you didn't go nowhere so you didn't get lost. There was always somebody around you could holler to or something. You know that you could stick your hand on your nose and you couldn't even see it?

This is not to say that all miners were sensible or even in agreement about the quality of their work environment, its dangers, and its drawbacks. Ask a group of miners about the smells in a mine and the answers will range from the stench of a rotting mule carcass or the odour of gas mixed with the stink of stagnant water, all the way to the smell of blackberry jam and marmalade, the two most popular sandwich fillings. Ask a group of miners about the sounds in a mine and the answers will range from it being too noisy to talk to it being so quiet they could hear the subtle creaks and rumbles that warned of danger. Ask a group of miners about the danger and this is what they'll say:

> I think it's against nature to work in a mine. So many things can happen. And there's only one way you can get out and if that hole is blocked, you never get out.

> It's dangerous sometimes but that's because people make it that way.

> The only thing I was afraid of was tapping an old mine and having a flood. I didn't want to drown. Wasn't afraid of cave-ins, just water.

> I never thought of the mine as a dark and eerie place. It was just like you were going to pick potatoes somewhere in a field. You just went to work. You grow up in it. In those days mining was all the work here except logging and that was nothing at all. I never saw anyone superstitious. They got up to work just like they were going to work in a store or bank or something.

> My mother mighta worried but it was just a natural thing in them days. Practically everyone worked in the mines; a lot of people never had cars and you got a job where you were living, something close to home and that was the mine.

The mine had always been close to home for most of Vancouver Island's coal mining families. Whether from Durham in England, Kilmarnock in Scotland, or Nantyglo in Wales, whether from Glace Bay, Nova Scotia or Beatty, Pennsylvania, the majority of coal mining families on the Island had been digging coal for generations. They were recruited first by the Hudson's Bay Company (HBC) and later by a succession of entrepreneurs whose need for miners on the coal-rich, labour-poor West Coast of Canada was never satisfied. They were lured by the promise of higher wages and better working conditions. They came in families and they came singly, from China, Italy, and Croatia. They came to better their lives. They came unaware that the mines were among the gassiest in the world, that unions were prohibited, and that once they were here, especially if they brought their families, they would never have enough money to leave.

> In 1909 there was an explosion in Extension coal mine and thirty-two men were killed. After the explosion the miners were afraid of the unsafe conditions of the mines and started to think about forming a gas committee and a union. Mr. Dunsmuir heard about this from his suckers and advertised for miners in Scotland and other countries as well, I guess. My father heard about this and thought it looked good so he came to make his fortune. He got here in 1909. My mother and all the rest of us kids got here late in 1910. Two years later when the Big Strike started, my dad still owed the CPR all the money for our fares.

The Canadian Pacific Railway (CPR) brought boats full of immigrants across the Atlantic and loaded them into colonist cars. The cars were bare with wooden benches and a heater at one end for cooking. Passengers had to buy their own food along the way, an extra expense and inconvenience few had known about in advance.

> My father came out in 1910 and worked over here for a while and then sent for Mother, who was still in Outerside in Cumberland. She brought six children out here on her own. We disembarked at Quebec City and took the route across the country. I was the middler, third in line, and I had to look after a younger brother and a younger sister, which was quite an undertaking on board a ship and on a train. I wasn't seasick, but I think Mother might have had two meals during the whole trip from Liverpool to Quebec.

It took two or three weeks to cross the Atlantic Ocean by ship and Canada by rail, but before the completion of the CPR in 1885 the trip to Vancouver Island took much longer. The earliest miners endured a six-month-long ocean voyage from Britain around Cape Horn. Later immigrants tried their luck in the United States, hopping across the continent from job to job until a recruiter in San Francisco or a letter from a relative lured them to the Island. Still others tried Australia first.

Political upheaval or poverty at home, utopian quests for the perfect place to live a perfect life – these reasons drew another kind of immigrant to the Island. Less likely to be experienced miners, Croatians, Italians, and Finns came singly or in groups, as bachelors or husbands with families, as complete strangers or to join uncles and brothers already here. These men spoke no English, knew nothing about the customs of Canada, and many bore an additional burden – knowingly or unknowingly they came as strikebreakers.

Chinese and Japanese men were also used as strikebreakers. Their situation was different, however. European immigrants might experience alienation or even contempt, but as they learned English they blended in. Asian immigrants came to a province full of racial hatred and were forever outsiders.

Racial discrimination in Canada has usually had economic roots. The newest group, easily identifiable because of skin colour, dress, or language, threatens the people already here, especially those nearer the bottom of the economic heap. Newcomers look different, they eat peculiar foods, they speak gibberish, and, worst of all, they work for lower wages.

For some racial groups in the mines the hatred died hard. The Chinese suffered discrimination long after they ceased to be an economic threat to other miners. The Italians who worked during strikes suffered long after the picket lines came down. The desire of Finns to live closely together drew suspicion from those outside their tight communities. Yugoslavs worked for labour contractors at half wages and earned themselves long-lasting resentment from the men they displaced. But down in the mine racial distinctions blurred.

> When they're down in the mine they were all friends. Outside, it was anybody's guess but down in the mine they're all brothers. They all want to see daylight. It didn't matter if you hated the guy outside, if he asked you for a hand down the mine you went right over and helped him.

> Even if they're not friendly on a personal basis there is a tremendous spirit of personal regard for other men and their safety. If they didn't see or hear the other miners around them for a while they would stop work and go and look. And they would call, "Are you all right?" If you had any heavy work to do, putting up timbers or stringers, you could always call on the other men around you and they would leave their work and come and help you. And of course, for a miner to leave his work was to leave off loading coal which is the basis of his earnings.

> We earned every nickel we got. When you went down the mine you just peeled right off to your undershirt and you'd see the steam coming off some of them fellers' backs. Sometimes you could smell the garlic comin' out of the pores of some of the Italian fellers.

> In Granby there was a great number of Italians. And they were good men. Because they're big men, strong men. And for that type of work they're a good worker. In the first place there is no Italian that I know that was raised without work. They tell me in Italy that if you don't work, you don't eat. And an Italian is a hell of a good worker. Now I'm English, but I'm not built like some of these Wops. Boy. They're skookum.

The underground camaraderie showed itself in the nicknames some of the men carried for life. It is not hard to picture what Skinny and Cheeks looked like, but Creep and Buff and Sledge are a little harder to visualize. Snore was a bad man to nap beside and Yacca probably dominated the conversation. Two no-hit baseball games earned Ironarm a lifelong title.

These men in their infinite variety populated the Island coal towns from Cumberland and Bevan to Lantzville, Wellington, Northfield,

Nanaimo, Extension, South Wellington, Granby, and Ladysmith. With their wives and children they formed a network of families and communities who came to the Island to better their own lives and stayed to develop the far western edge of a young country.

Long before that country became a nation in 1867, decades before the building of the CPR, when the prairies were wide open and barely touched by Europeans, when the Rocky Mountains were an exotic barrier known in eastern Canada only through the paintings of Paul Kane, there were coal towns on Vancouver Island.

The HBC had not been looking for coal when Chief Factor James Douglas chose the southern tip of the Island as the site for Fort Victoria in 1843. The company, with its vast monopoly encompassing all the territory west of Ontario, including what is now Washington and part of Oregon, was in the fur-trading business, had been for two hundred years. Douglas had agreed to attempt colonization, but there were no plans to even look for coal in this isolated place that was a six-month voyage away from Britain on treacherous seas.

But Britain was the leading sea power in the world, and her navy was gradually converting its ships from sail to steam. Steamships needed coal in various locations all around the world to fuel their boilers. It was with some interest then, that Douglas received word, in 1852, from the company blacksmith that a Sne ney mux man had reported seeing coal on the beach at Winthuysen Inlet ninety-six kilometres up the Island from Fort Victoria. An earlier find at Fort Rupert on the northern end of the Island was proving too difficult to mine. And the Scottish miners who had come to work the coal in 1849 had turned out to be a difficult lot.

The report of the Sne ney mux man, known later as Coal Tyee, resulted in the establishment of Nanaimo on the shores of a natural harbour, which would soon be filled with naval vessels, and later with colliers bringing ballast rock from all over the world in exchange for the black fuel that was becoming as necessary to the world's economy as breath was to life.

The HBC was not successful in the coal business. By 1862, it had sold its operations at Nanaimo to the Vancouver Coal Mining and Land Company (Nanaimo Coal Company), a British stock holding company which took over responsibility for the mines, buildings, a generous land grant and the livelihoods of the inhabitants who had been recruited from Britain – the men for their skills, the families for their stabilizing influence.

This was an adventure of immense proportions to men reared in the narrow confines of British coal towns. The company feared that the men would become discouraged when faced with the isolation, the climate, and the working conditions on the rain-soaked island so far from anywhere. Wives and children would offer much comfort to such men. And, more important from the company's standpoint, men with families would find it more difficult to leave if they became dissatisfied.

And there was much room for dissatisfaction, especially when a second company began to operate. Dunsmuir, Diggle and Company began to mine coal after Robert Dunsmuir discovered the black treasure near Divers Lake just north and west of the Nanaimo Coal Company land grant. A Scottish miner brought to Fort Rupert in 1851 and later to Nanaimo by the HBC, Dunsmuir had been looking for his own coal seam since 1853. When he stumbled over the outcrop of a coal seam, he was already forty-four years old, but in the twenty years before his death in 1889, Dunsmuir established an empire based on coal from his mines at Wellington, and on a huge land grant he received from the Canadian government as partial payment for building the Esquimalt and Nanaimo Railway (E & N). He also acquired a reputation for ruthless labour policies, which helped contribute to his burgeoning wealth and which left thousands of his employees feeling oppressed and fighting mad.

Dunsmuir's son and successor, James, had even more political influence than his father. As premier and later lieutenant-governor of British Columbia (B.C.), he added lustre to a growing legend, a legend enhanced by his daughters' brilliant marriages to European aristocrats. The Dunsmuir legacy, however, foundered on a lack of male heirs. By 1910 the coal interests had been sold to Canadian Collieries, and by the 1930s most of the fortune had disappeared. What remains of the Dunsmuir legend, in the coal towns of Vancouver Island at least, is a legacy of oppression. The name is synonymous among coal miners with the tyranny of the boss.

The boss gave the orders; the boss owned the mine; the boss hired lesser bosses who in turn watched the men. There were fire bosses, pit bosses, shift bosses, boss bosses. There were some good bosses and a lot of bad bosses. The most tyrannical of all bosses was the boss whistle. It told a miner when to come to work and when to go home. Its strident voice howled when a doctor was needed to help an injured man, and it roared continuously when an explosion took the lives of many. The boss whistle was silent when the Big Strike closed the mines; it bided its time and then sounded again defiantly as it called strikebreakers to work in the place of union men. When the coal markets began to die, the boss whistle assumed even greater powers. Families listened each day for its voice, their livelihood depending on its message. One whistle: work tomorrow. Two whistles: another day without work. And when a retired miner stopped on the street to check his watch with the twelve-noon whistle from the mine, he was acknowledging the lifelong influence of the Boss Whistle.

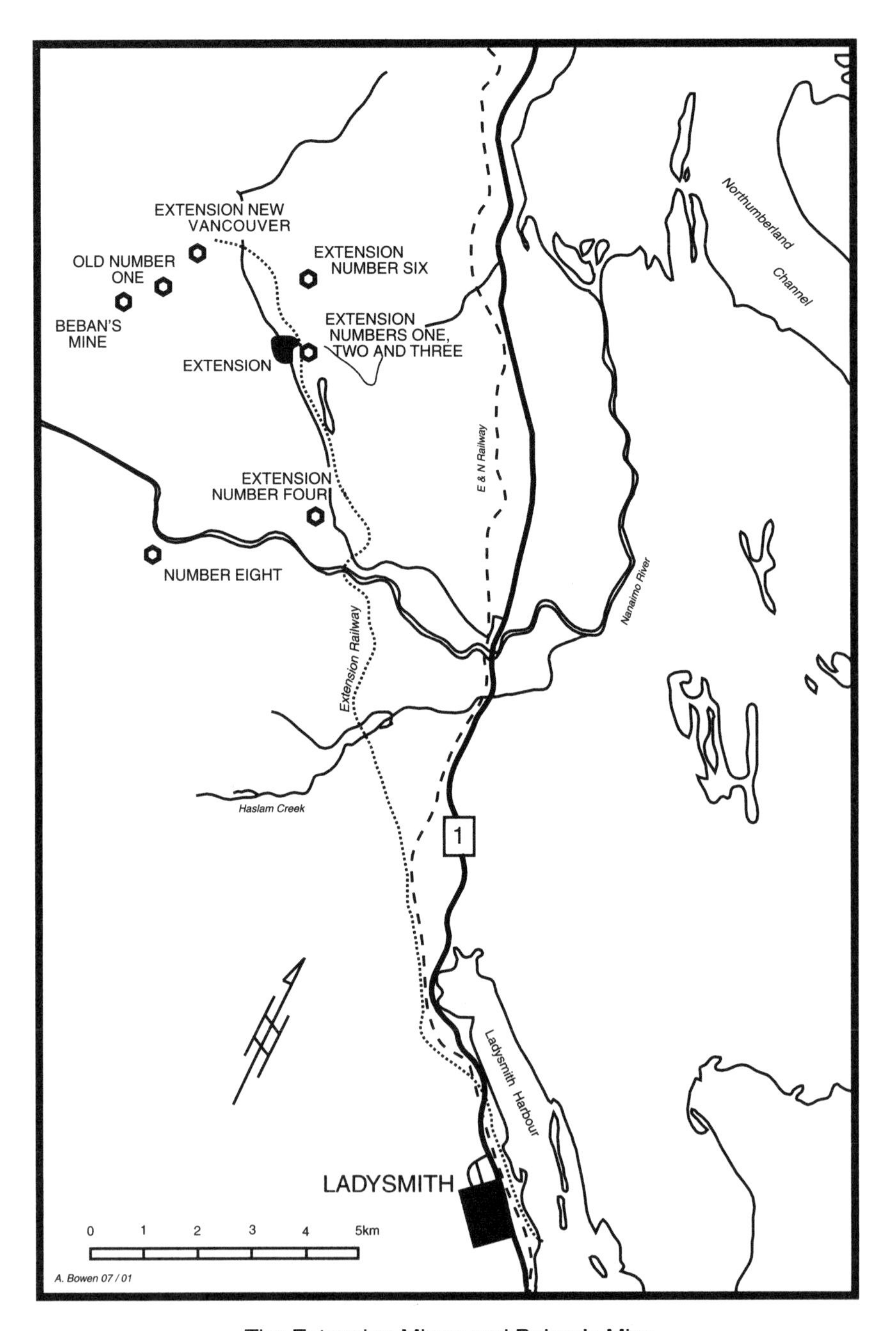

The Extension Mines and Beban's Mine

CHAPTER II

The Put-Up and Pull-Down-Again Towns

> Oh Extension is the Devil's Hole. It was no place in my day, Extension wasn't. I think someone called it Devil's Gulley.– Mrs. Frank Wall

> It was the end of the world to me. When you come from a city in England to an old mining camp and all you see around you are a bunch of stumps and rocks, it's like the end of the world. – Albert Steele

When John Bryden saw the homely little coal camp of Extension shortly after it sprang into existence in 1895, he thought it "a most undesirable place. The whole camp seems to be built in higgledy-piggledy manner," said the former manager of the Wellington Colliery and son-in-law of Robert Dunsmuir, who had died in 1889 leaving control of his empire to his wife and two sons, James and Alex. Six years later, just when the Wellington pits were showing signs of depletion, a Dunsmuir employee, one Ephraim Edward Hodgson, found coal clinging to the roots of an upturned tree at a location south of Nanaimo within the E & N land grant.

The land, complete with mineral and timber rights, already belonged to the Dunsmuirs, and since the coal Hodgson found came from an extension of the Wellington seam, the company named the new coal camp Wellington Extension, soon shortened to Extension.

The owners of hotels and houses in the now dying town of Wellington dismantled their buildings, loaded them onto railway flatcars and moved them south. But hardly was the camp established near the new mine tunnel than James Dunsmuir decreed that the town would move to a site more advantageous for mining. The owners dismantled the shacks and hotels and rebuilt them on the new site. Then Dunsmuir decreed that they would move again, this time to Oyster Bay Harbour nineteen kilometres to the south. The official reason given for the move was the lack of water, and this was certainly a problem, but the people who had to spend $200 each time to move their buildings knew the real reason. The camp sat on land that didn't belong to James Dunsmuir. Hodgson's father-in-law, Jonathon Bramley, owned the property and rented building lots to mining families for $1 per month.

Bramley had bought the land from the Dunsmuirs in 1884 for $1 per acre (2.5 hectares). Now that there was coal to be mined, James Dunsmuir wanted to buy the land back, but although Bramley agreed to sell him twenty hectares for the railway yard and colliery, the two men could not agree on a price for the 6.5-hectare campsite.

Dunsmuir said the water was scarce and undrinkable; Bryden said he feared an outbreak of typhoid fever. But in testimony before a royal commission the two men confessed that they disliked having the miners living too close to the mine because it made them difficult to discipline. As Dunsmuir put it, "In the first place we were too near Nanaimo; that was the reason of a lot of trouble between our workmen. In the next place, it was not a fit place to live. . . Another thing. There was no water in Extension."

Although nobody ever saw the order in writing, it might as well have been: "Move the camp or lose your job." James Pritchard, who'd been thirty-four years a miner, said it was just understood that if they didn't move to Dunsmuir's new town, there'd be no work for them. A pit boss named Sharp told the miners, "I have an order that the men who do not come up on the train do not need to work." And James Dunsmuir said, "I told them they could live where they liked, but I would hire them where I liked."

Dunsmuir wanted them to live in the new town he was building at Oyster Bay Harbour; he would connect it to the mine with a branch of the E & N Railway, and he would call it Ladysmith in honour of a town in South Africa. The soldiers of the British Empire were at that time engaged in fighting the Boer War, and things had not been going well for them. When the British army finally relieved a siege on the town of Ladysmith in February of 1900, there was great rejoicing among British subjects all over the world, not the least of whom was James Dunsmuir, better equipped than most to make the grand gesture – no matter what the citizens of Oyster Bay Harbour might think. Many citizens of Extension decided that, edict or no edict, they preferred to stay where they were.

> "Take your houses with you and down to Ladysmith you must go." Well, some of them stayed in Extension but Dad and the majority of them went to Ladysmith. He built a good house, two-storey with five bedrooms. Had to with nine kids. But sometimes I could just cry thinking of the people that still live in them little houses that was built in the nineteenth century in Extension.

> It was cheaper to live in Extension – not so many city temptations.

> And you'd go up a hill and you'd look down at Extension. It was a hollow like and it was all kinds of shacks there. It might have been all Chinamen's

buildings for what you know lookin' at 'em. They were rough, you know, built rough 'cause they were built to put up and pull down again.

At first the people who stayed in Extension had to be secretive and pretend they had moved. They would walk down the railway track towards Ladysmith and catch the miners' train as it brought miners into Extension for their shift. Eventually, however, Dunsmuir chose to ignore them and allowed the town of Extension to exist.

The town that refused to be moved nestled in a green valley snug against the south shoulder of Mount Benson. Had there not been coal under the ground, it would have been a pretty farming community. But getting coal out of the ground left the town a dusty place noted mainly for the stumps left behind when the trees became mine timbers.

Extension was just like the pictures you see of a little mining town with little black houses where the soot comes all round, just these little black houses and people having a little bit of a garden and that's all. At first there were five or six hotels there but later all I can remember is the Tunnel Hotel.

And the roads in Extension were just laid around big boulders. Oh they were terrible the roads were.

What was Extension like in those days? Oh lots of fun. There was big Slav town and Italian town and Scotch town, all different names. Whatever nationality you were, you mostly stayed together. Chinatown was the main thing. They had street with stumps, oh boy, the stumps in there.

Stumps and miners were what Ladysmith and Extension had in common. Ladysmith should have been more genteel and picturesque, separated as it was from the befouling influence of the mines by nineteen kilometres of countryside. The haste in which it was assembled, however, the portable nature of its houses and hotels, the swampy ground, and the practice in those times of handsaws and horse haulage of leaving a one-metre stump when cutting down a tree, made Ladysmith just as much of an eyesore in its early days as Extension was. There were two-metre-high wooden sidewalks with swamp below and stumps beside, there was mud in wet weather, dust in dry and stumps all year round. As the town grew, an advance guard of stumps preceded the new buildings as they marched higher and higher up the steep hill.

When Dunsmuir first decided to move Extension to Ladysmith, a lot of the houses were just torn down piece by piece and transported down there to be rebuilt again, exactly in the same shape and style as they were torn down. And we used to have building bees. That was the common thing especially in Finn Town. Everybody building at the same time. And there was always shingling bees around. As we grew older us kids used to have to go up on the roof and help the men. We laid shingles and they nailed them on and even when we were seven or eight years old we used to lay shingles as good as the men anyway. The women always brought sandwiches and coffee.

Except for rough trails, train tracks were all that connected Ladysmith and Extension with the rest of the world. Dunsmuir had at first intended to ship Extension coal at his wharves at Departure Bay north of Nanaimo and had begun to build a rail line for this purpose. But the Dunsmuirs' poor relationship with Samuel Robins, the general manager of the Nanaimo Coal Company, got in the way. In order for tracks to reach Departure Bay from Extension it would have been necessary to cross Nanaimo Coal Company land. Robins' refusal to grant permission forced Dunsmuir to build shipping facilities in Ladysmith that connected to Extension with a separate railway line.

In addition to transporting the coal to tidewater, the track acted as a two-way umbilical cord between the stores in Ladysmith and the residents of Extension and between the mines in Extension with the miners living in Ladysmith. Steam-driven locomotives pulled three passenger coaches, which were heated with coal-burning pot-bellied stoves. Shutters, not glass, covered the windows, and provided only marginal protection from the cinders that flew from the engine's smokestack. The ride was free, but most people called the coaches cattle cars.

Three times a day the miners' train made the round trip between Ladysmith and Extension, tooting its whistle to announce its arrival and departure. Before 1917 when the company built washhouses at the mine, the returning miners were black as well as tired and wet. But their children met many of them at the station and escorted them home to a hot bath prepared by the women, who had pumped the water, heated it on the kitchen stove and, when they heard the train whistle, poured it into a galvanized tub sitting in the middle of the kitchen floor. One tub per household. If a house had more than one miner, the first one home got the cleanest, hottest water.

Homes in Extension were just as primitive as those in Ladysmith and remained that way for longer. Even in the 1920s, houses had only coal oil lanterns and outdoor toilets. In the summer the communal wells produced little water, and only the first in line got some each day. With only one hotel and a small store left after the exodus to Ladysmith, the residents depended on outsiders for the goods they needed. Merchants from Nanaimo, forty-five minutes away by horse-drawn express cart, brought goods right to the customer; clerks from each of the bigger stores in Ladysmith came up by train to take people's orders, which the train delivered in due time. Families bought on credit and paid monthly, the storekeepers rewarding the children with a bag of candy on payday when their parents paid the bill.

Extension children went to a one-room government school built on a rocky bluff almost a kilometre from the coal camp and cut off from it by an E & N fence. It was easier to jump the fence than walk more than two kilometres around it. The teacher presented another problem. He was a tippler, and the children knew he was going to go out back for a quick snort if he told them to study nature or paint. While he was gone, the older boys jumped out the window, leapt the fence and headed for the woods.

But their families were large and living conditions primitive, and even the most mischievous boys went home when their mothers needed them.

> If my mother wanted anything from the store in Ladysmith, she had a note for me and as soon as I came home, I'd catch the train which got into Ladysmith at 4:00 and it left at quarter after four. There was a store right close to the station so I'd have to race in and get what she wanted and get back to the train in time.

On the morning of October 5, 1909, anyone in Ladysmith who glanced down at the train station would have seen an unusual sight: the mine doctor, followed by several mine officials, boarding a single rail car behind a locomotive. The train left the station heading for Extension, but no explanation was forthcoming until late that afternoon when news reached Ladysmith that there had been an explosion in the Extension mines.

> My mother was only here just a week when that happened. She nearly went nuts. This fella comes into the store and he's tellin' my uncle something in English and she seen him turn pale. Right away she knew there was somethin' wrong. She kept buggin' him; he wouldn't tell her. Finally he said, "Well, there's an accident in Extension." But my father was all right and that calmed her down. He was lucky he got out. He and his partner were timbering and they went to the other side of the trap door that divided the air. They were packin' this heavy six-foot post. The explosion blew that door to smithereens and knocked him down on the track. My dad got a cut on his head where he hit the rail. Everyone on the other side of the door got killed.

At two and a half west level, the fire boss had checked for gas at 6:00 a.m. just before the beginning of the day shift. As it had been for the past two months, the gas level was 2 percent – not high enough to make naked flame lamps dangerous. The fire boss marked each place with chalk – signifying it safe – and moved on. A gas level of 3.25 percent would have closed the mine temporarily. A level of 3 percent would have required exclusive use of safety lamps, but since Extension mine was an open flame mine – the miners lit their work with oil lamps which burned with an unprotected flame – such a requirement would have shut down the mine as well because there were not enough safety lamps available. But the gas that day was only 2 percent and that was only dangerous if a runaway car caused a spark, if a blasting shot blew outward instead of inward or if a cave-in raised coal dust.

Extension mine was a wet one, which should have meant that the coal dust was under control, but when a section of roof caved in at 8:30 a.m., the combination of the dust, the gas, and an open flame ignited an explosion. Doors blew from their hinges, rails heaved, posts flattened, and stringers crashed to the floor. All over the vast mine, miners knew from the rush of air, the swinging aside of brattice and the dull rumble that there had been an explosion. Experienced miners knew that the only safe place was outside the mine in the open air. Carbon monoxide or afterdamp, an odourless lethal gas, was even then spreading throughout the mine. The fortunate would outrun it. The gas would overtake the rest and they would die.

Two men were loading stringers. One shouted, "By God, she's blasted." A rush of wind blew their lights out and nearly swept them off their feet as dust and dirt swirled round them. Not daring to light their lamps again, they were glad when nine other miners, two with safety lamps, joined them. The eleven groped their way along the level. Hot stifling air burned their throats, smoke and dust obscured their vision, and they were unsure where the afterdamp was lurking. Fred Ingham, the man in the lead, began to feel weak, and called out to those behind him to turn back. Six men were able to retrace their steps until they found air they could breathe; they decided to wait there and gamble that rescuers would reach them before the gas did. The gamble paid off and they made it to safety; the rescue team found the other five men. They had lain down close to each other before they died.

Tom O'Connell, star fullback for the Ladysmith football team, had called to a friend after the explosion. Then he charged away towards the good air, his athlete's legs propelling him in a headlong dash until he collided with a pillar in the darkness and the gas caught up with him.

The news reached Nanaimo by 11:00 a.m., and reporters and the curious hurried out to Extension by bicycle and rig. A crowd of onlookers soon surrounded the entrance to the mine at the base of the hill. It was not an unusual scene to people living in coal camps. The sight of a sombre crowd clustered near the entrance to a mine was a common one. This crowd was different only in that it contained less than the usual number of women. No one had informed the town of Ladysmith, nor would they until late in the afternoon.

The *Nanaimo Free Press,* October 6, 1909

To those who have visited Extension, it need not be said how dreary and cheerless is the outlook on a dull October day. Everything was intensified yesterday. A crowd had gathered about the mouth of the tunnel which

> shoots right away under the base of the hill. The mouth of the tunnel looms dark and damp and uninviting and somewhere within its depths lay the tragedy of death with brave men toiling away in noisome fumes to reach their imprisoned comrades.

Without the oxygen apparatus that would make such operations easier in later years, rescuers laboured to restore the ventilation system that would sweep the afterdamp away and allow them to reach the site of the explosion. They saw the first body at 4:00 p.m. The explosion had thrown Alex Milos, a gigantic Greek known for his feats of strength, seventy feet, fracturing his skull and his right leg, and shredding the skin on his bare chest. But he had still been alive and running from the afterdamp when it caught up with him. They lifted his body onto the top of a loaded coal car.

At 5:00 p.m. the crowd outside heard the hum of the motor. It appeared at the portal of the mine tunnel pulling five cars. In each car was a dead man, but of the five only Milos showed any sign of physical injury. Each man, his number chalked on a shoe or overall, was carried up to the second floor of a large storehouse and laid beside his comrades. When all the bodies were finally in the storehouse, the authorities allowed relatives in to identify all the bodies. Seventeen were burned and bloody, but fifteen showed no injuries whatsoever. The afterdamp had killed them.

Tom O'Connell was twenty-seven. His wife's brother, Robert White, was forty-six. Fred Ingham was twenty-three and the third of his mother's sons to die in mining accidents. Eddie Dunn was seventeen. Jim Molyneaux, who could sing so sweetly, was married with children. Thirty-two men died that day, thirteen of them married; thirty-eight children lost their fathers.

They say mining is a dangerous occupation, but the thirty-two did not have to die. The inquest found that the unlikely combination of a poorly set shot, a little gas, a certain amount of coal dust, and an open flame lamp had caused the explosion. The company had gambled that in a wet mine with low gas levels, there was no need for every man to have a safety lamp. The company had gambled that the men knew enough about blasting to fire their own shots when the conditions were deemed safe. This increased efficiency and saved the expense of hiring more shotlighters. But the company was using men's lives as currency, and the gamble had not paid off.

The inquest, however, absolved the company of all responsibility. It found nothing reprehensible in the fact that Robert Dunsmuir and Sons persisted in using open flame lamps when other local mines used safety lamps exclusively.

In the company storehouse, volunteers wrapped each victim in canvas, pinned a ticket bearing the name to the shroud, and placed the body in the coal train caboose. Two bodies went to Nanaimo. Thirty went to Ladysmith to their homes or to the Finn Hall.

> I went up to the mine to see them bringing out the dead people. 'Course you couldn't see some of them 'cause they were all burnt black you know. Yeh, the men was all burnt, killed in the explosion. And the mules and everything. 'Course you know how kids are. They want to go and see what's doing.

> I would be about five years old. I remember going into the big hall the Finnish people had and the coffins—I can still see in my head the coffins stretched out in the hall.

There were thirty-two funerals between October 7th and October 10th, all but two of them in Ladysmith. Almost all of the adult population watched as one after another, the coffins – some on the shoulders of lodge members, some in hearses – were carried up the hill to the little graveyard in back of the town.

> I've been in the graveyard there several times and seen a lot of gravestones of the people who were killed. The funerals were mostly on foot. Ladysmith had a band and it played for every funeral. They walked slow and they went and picked up one and took him to the grave, buried 'em and then went back and picked up the next one. It was at least two days. I still know the tune they played. It's kind of eerie. Sometimes it rings in my ears even yet.

Thirty new graves in one cemetery is a sight not soon forgotten. Many men felt that the time had come to fight back. Two years earlier, some of the men had even formed a miners' committee to check for gas, but had given up when the company resisted. The 1909 explosion, however, made a lot of men question how much longer this could go on. The widow of one of the dead men sued the company for negligence and lost; but when she appealed, the company offered an out-of-court settlement of $1,500, which convinced the mining community that the company had a guilty conscience.

As Ladysmith and Extension began to adjust to the loss of so many of their fellow citizens, as the widows cast about for ways to support their children after the relief funds ran out, even as angry thoughts and words spurred men to think of organized resistance, the Extension mines reopened. Eight days after thirty-two men had died, the living went back to work.

The modern visitor to Extension at first sees no sign of where the men worked. The road comes over a hill and a small green valley lies below. The stumps have rotted away and trees have planted themselves and grown up; houses occupy the streets in random fashion, spaces marking where their predecessors were and are no longer. Some of the surviving houses are dignified in old age; others are disreputable. Newer houses seem out of place.

The road dips down to the valley bottom and then begins to wind up again toward a house that used to be the general store. A trail leads back to the valley floor and an immense pile of black rock looms through the trees. This and a low concrete arch tucked into the hillside beneath the former general store are all that remain of the Extension pithead.

Where once double tracks carried small electric locomotives or motors that pulled coal trains out of the tunnel, now a stream spills out of a low silt-filled opening. Nature has almost reclaimed the area, has camouflaged the entrance with trees big enough to tempt a logger. But to those who worked here and to the ones who waited at the entrance for news after an accident, this place is alive with memories.

> There was lots of noise here when the mines were goin'. Locomotives chuggin' back and forth, mine bells ringin'. Steam plant runnin' down at the mine makin' quite a bit of noise. And the coal getting dumped into the cars. And at the same time that there rock dump was all burnin'. It burnt for years. Little fires all over it. So when you entered this valley, all you could smell was sort of a sulfury coal. There was some coal mixed with the rock and that was what was burnin'.

> The rock dump was started where the old manager's house used to be before the strike. It was burned down during the riot in 1913. Some of the apple trees from the manager's orchard are still there.

> Everyone called that rock dump Mount Bickerton in the '20s because Mr. Bickerton was the outside foreman for the pithead. It was the biggest mountain around here. There was a track going up it and the rock from the mine would be dumped into special steel cars, sort of wedge-shaped things that went up the steep incline to the top. There was a little shack up there in the early days and two Chinamen stayed up there and opened the doors on the cars and dumped them. Later on the cars dumped automatically.

In the shadow of Mount Bickerton young boys – too young at thirteen and fourteen years old to be underground – loaded timbers and cordwood onto special flatcars, tended mules, pushed empty cars and assembled them into trips to be returned to the mine, and filled powder cans in the lamp house.

> They had a room in the lamp house where the miners put their powder cans and we filled them. And they handed them to the guys that's goin' into the mine. It was all done by check number, on the lamp, the powder can and their tools. They had to pay for their powder and the lamp too.

After 1917, when mining went electric, miners no longer used fish oil lamps and safety lamps continued to be used mainly by fire bosses to check the mine air for gas.

I serviced the lamps, charged the batteries and got them ready for the shift comin' on. These were electric lamps. I don't remember fish oil lamps. The ones we used had a battery strapped to the belt and the cord went down from the hat to the battery.

There were no lights in the mine except for the one on your head. You had a battery weighed about two or three pounds. Every shift you'd put them in the lamp house and they'd put them on the charger. They'd last a whole shift, oh sure. That's all you had for light except some of the rope riders were allowed spotlights. Those had reflectors on 'em and they blinded everybody else. The rest of us just had flare lights, like the light sprayed out. So if you heard anything, a rat or something, you couldn't spot it with your light.

When the pithead work was done, the boys would persuade the motormen to let them ride into the mine. That was where the real action was, and that was where they longed to be.

I think there used to be about a mile straight you could see lights. I think it was about a mile straight in from the pithead, then it branched off to three different mines. Number One went off to the right, Number Two to the left down towards the Nanaimo River and Number Three was further in to the left and up towards Mount Benson. It took in quite an area, altogether around seven miles.

When you're in the mine, you branch off, some levels go up and some go down. Because the coal seam isn't level. Used to be at a little bit of a pitch. And they had a railroad going in about the middle of the mine and then each level had a name or a number. And then, even the place – they called 'em a room or a stall – they call that by number too. So you know if something happened, they know where each place is. Or when the fire boss goes and inspects each room every night before the men go in, well you just ask him a certain number. Is it all right?

'Course Extension was high seam, up to fifteen feet high. With high seam you use pillar and stall. Everything's blocked off like a city. You have your first avenue and your second avenue and then you have your streets going up and then you have your alleyways and your crosscuts to conduct air and to get around the places where the men are working.

Now this is what I mean by pillar and stall. You drive tunnels maybe seven or eight feet high, twelve feet wide parallel to each other and then you drive a crosscut every so often so it's like an "H". What's left is the pillar and you leave that until you're working back out when the mine's finished. These pillars are maybe forty, fifty feet square and that's solid.

Two or three miles from where the three mines branched off the main tunnel was the face where the miners dug the coal. Drilling and digging by hand – there were no air drills in these mines – two men in a good place could fill fifteen mine cars a day. Each car in the Extension mines held between 1,170 and 1260 kilograms of coal. In mines owned by the

Dunsmuirs, men usually worked alone with an occasional visit from the fire boss. If a miner injured himself, or needed help to wrestle a timber into place, he hoped a nearby miner could hear him when he called. Most miners were glad to help, even though time away from their own places meant lost revenue. The Nanaimo Coal Company mines, now owned by the Western Fuel Company, required two men in each place, but in the Dunsmuir mines there was no such rule.

The young boys riding with the motorman knew all about the mines from listening to the talk at home. They heard as much about the dangers of mining as they did about the camaraderie between miners, but most of them still wanted to be diggers.

If a boy who lived in a coal camp got tired of school and was anxious to make a little money, the obvious thing for him to do was to go to work in the mines. If he was willing to lie a little and add a couple of years to his age, he could start work on the surface at age thirteen or fourteen.

For the first year or two he would work at the pithead, with the Chinese on the picking tables for his workmates. If he were lucky, however, he would soon go underground. That was where the money was if you were a digger and on contract. A boy could not become a digger right away, however; he had to work his way up and learn mining techniques so he could pass the examination for his miner's ticket. So first he ran a winch hoisting loaded cars up steep inclines, and then he became a rope rider, tending a trip of loaded cars as the haulage system pulled them from the face to the shaft bottom.

From rope riding many went to driving a mule. The mule and his driver took empty cars from the bottom of the last hoist and delivered them to the diggers. The diggers filled the cars and hollered for more. There was constant pressure on the driver to supply empty cars and take away filled ones. A digger was only as good as the number of empty cars he could fill. If a driver was unable to keep the cars coming, the diggers cursed him. He in turn cursed fate or the "doggone mule" when the animal balked or when a heavy car derailed.

> It's hard to explain the exasperation. I saw one fellow get down on his knees and pray to the Lord to strike him dead. See, the boss is after you if you don't get the coal out. And everything is against you. You're going along good and then a car goes off the track. Well, there's nobody near so you're on your own. I've broken a lot of laggin' trying to get cars back on the track. You get the car so it's just about back on the rail and all you need's a little more sloo, then it would fall back down. We used to say, "God, oh God, if you be true, lift this car and I'll sloo."

Every mine that used mules for haulage spawned a host of stories about the "doggone mules." More than mere beasts of burden, the mules assumed personalities and gained the drivers' respect for their intelligence. In an area too low for the mule to stand upright, the animal would go down on its knees to get through. On steep inclines, they learned to push the cars with their chests. Pursued by a runaway car, a mule would duck into the first recess in the wall and let the runaway go clattering by. And the mule knew when it was quitting time.

> You'd try and chase them in there for an extra trip and they'd balk. They knew just when to come home. You just threw the harness off and tell 'em to go and away they'd go to beat the band. Sometimes we'd grab a hold of their tails and slide along the rail.

Driving on haulage was tough and dangerous work. The driver had to be alert both to the dangers of the mine and the mule. He could be caught between moving cars or his hands squeezed between the top of the car and a low roof. Pushing the car the last few feet after he unhitched the mule was very heavy work. The driver and the mule had to be a team. They had to declare a truce in the state of war that existed between them, especially when the driver's feeble light went out from a sudden jolt. Then both were dependent on the animal's remarkable sense of direction, which would lead them back to the stable in blackness so intense a man could not see his own finger in front of his nose.

Mules were smart, but mules were ornery. They responded to a certain man whose touch they trusted but could just as easily refuse to work for him if the load was heavier than usual. A stubborn mule could drive the most compassionate man to tears of frustration or to smacking the mule across the rear end. The mule did not let that go unrewarded either.

> Oh there's some mean mules, boy. They'd squirt you and some would bite. As you went by them they'd snap your arm if you weren't watchin'. And kick? You'd harness them and they'd have a chain about six feet long with a hook on it to hook cars. They'd keep kickin' that chain until it hit you in the head.

> I drove this big Blacko one day. He was kind of an erratic mule. In fact when I was off work nobody else could drive him. He'd just stand there and shake his head. And he wouldn't back up very good and I was pulling on the tail chain and he let drive right between my legs. Just about got me.

> I drove one by the name of Major. We had to push him into the mine. One morning I put my hand on his rump and oh boy, he let me have it. He got me between my legs and he lifted me oh I guess twenty feet. Sailed me right through the air.

> Oh yes, they'll drive you crazy. No wonder a guy can swear. One thing I used to wonder, how can a man swear so much? That's how a lot of foreigners learned English – listenin' to the swearin' in the mine.

In the perpetual war between man and mule there were a few casualties. The mule was smart and quick with its heels, and the mine owners seemed to put a higher value on a mule's life than on a man's, but some men and some companies, deliberately or inadvertently, were cruel to the animals.

> In some cases the roof was low so the mule had to scrape himself through and they got their backs all scratched. If they had a good animal they would abuse him. They kept him in there even twenty-four hours instead of changing him every eight hours.

> There was a hundred mules in the three Extension mines. The mules very seldom got old. Somebody killed them. Some people treated them cruelly. One day the manager told me to take twenty mules up on a bluff away from the mine. You see the government inspector was comin'. The manager says, "Keep watchin' the office winda upstairs and when you see me wavin' a white flag, bring them down." They was all raw. Some of them had no hair on their backs at all, just red flesh. I was up there about four or five hours I guess. I kept mindin' the winda, then I see this tablecloth or somethin' and I took them back to the barn.

The abuse of mules in coal mines may seem cruel, but mules were far better suited to hauling coal cars than were the women and children used for that purpose in British coal mines before 1850. Mules, with their short legs and long bodies, were well suited to mining. What appeared to be stubbornness was really an inherent prudence that made them stop when they had worked too long or when they sensed danger. All they needed was a patient and tolerant muleskinner to get the most out of them. Such a man, so the saying went, thought like a mule.

Old-timers remember Maude and Pixie and Jerry, the choice of names ignoring the mules' lack of sexual identity. They also remember the worst mule around. Liza was so mean it even kicked at its own reflection in the big windows of the office building. The mules of Extension mines lived on the surface unlike those of many other mines whose stables were underground. Mules had to be washed daily, the water and mud being a constant problem in Extension for both mule and man.

Because the tunnels of Extension mine angled slightly upward, the underground water ran downhill and eventually out through the entrance.

> There was about two feet of water in the big main tunnel all the time in the winter. It was running out like a river, drippin' off the roof. Your feet were never dry. It would dry up in the spring and start up again in the fall.

> I've worked in the wet quite a few times. Some places, if you had swamp on top of the ground, well it seems to come right through into the mine. Even very deep under the ground, the water seems to come through. And some

places even if there is no water on top, there is different streams in the ground. It might be dry on the surface and then down so many feet it might be a stream.

In the wet tunnels and crosscuts, the diggers drilled and blasted the coal; they loaded it into cars, and the mules pulled the cars to where they were hitched to the motor. In the early days of the Island mines, the motor or "locie" had two generators, one for each set of wheels, powered by electricity from steam boilers tended by Chinese workmen. Later, a powerhouse provided electricity. The little locies pulled up to 120 cars at a time over a miniature version of the system of overhead wires used for trolley buses.

When the cars got outside to the pithead, well then they dumped one at a time into a chute. You run them on to a level kind of a table and when they got up to the front, the weight tipped them over. The front door of the small car opened, a hook grabbed the door to keep it open, and the coal run down. And so this coal would go down the chute and dump into the bigger railway cars which carried forty tons each. The coal would have been screened going down and the slack coal would go into one big railroad car and the lump coal would keep on going on to what they called the sorting table or picking table.

There was a lot of Chinese on the picking table throwing rock out of the coal and then this lump coal would be dumped into another big car. When the car was full, it would go ahead and another empty come and it was on enough of a slope that they'd pretty well run themselves down there with one chap just brakin' 'em. Then a big locomotive would haul the train down to Ladysmith harbour and those cars would be dumped into boats or scows there for to be hauled away. The washery was down there on the docks. They washed the slack coal to remove small rocks and screened it to get the different sizes separated.

The classifications of coal were lump, egg, nut, pea and slack, although other markets use other descriptive names. In Vancouver and Victoria at least seventy-six different descriptions of coal were available.

Digging, grading and transporting the coal employed many workers in the crash and rumble of the tunnels, pithead and docks, the airborne dust adding a sinister atmosphere to the scene. And often it was a scene of tragedy. Between 1899 and 1931, the mines at Extension killed 104 men.

Seventeen men died on September 30, 1901 when spontaneous combustion started a fire in "Old Number One," the original mine, whose entrance was above and behind the main Extension tunnel. A burning coal mine feeds on itself. Rescuers can become killers when they are forced to block the entrances and exits to cut off the oxygen supply to the fire. Any men left alive inside will die in the flames or from a lack of oxygen or by drowning when the mine is intentionally flooded to finally extinguish the fire.

> I was only a schoolboy and I was up on the hill above Extension near the big fan shaft for Old Number One and I seen the smoke comin' out of there and I ran home and I says, "There's lots of black smoke comin' out of the fan shaft." It was from the fire in the mine.

Very few people remember that fire but the story of the miner who escaped with severe burns has become part of Extension lore. It was a time when only those without family to care for them went to hospital. This miner was nursed at home, his family laying him on wet sheets so they could turn him over more easily.

Large explosions and fires with multiple fatalities tend to make the newspapers, but mining killed men one at a time as well. In the cemetery at Ladysmith are the graves of miners killed by a fall of rock, a fall of coal, a runaway coal car. Some of the graves hold bodies of boys aged ten, twelve, thirteen and fourteen.

> In Extension for months and months they killed a man a month. It was a bad roof in the mine and quite a few fellows got their backs broke in cave-ins. I used to go and see two old fellows here. They were in bed for the rest of their lives. The formation is what got most of them – what they call hogbacks. A little dip. And there's a little bit of water in there so when you pounded it with your pick it sounded solid and that's what got most of them. It'd loosen up and drop in like a huge rock.

> Extension mine wasn't much on safety. And usually if a man got killed or that, there weren't much said about it. There weren't compensation like we get now. Once they had to start paying people that got hurt or killed, then they became a little more safety conscious. But at the same time, if a horse or a mule got killed down there, somebody got fired for it. They were more important than the men because they cost more. They could get men for nothing.

> Well when I was a kid my mother would be after me. "You look for a job. Go to Extension." I didn't want to go to Extension, but I got on my bike and went up there. I got there around 3:00 and the miners were comin' out of the mine. Well the first motor that come I see my friend's old man on top of the load, dead. I never asked for that job. In them days a man was worth nothin'. They never lost a car of coal. They just put him up on top of the load and took him out.

> My dad lived through the 1909 explosion but in 1929 he got killed. Two days before Christmas. Yeah. A cave-in. They were drawing pillars. It was a few minutes before quittin' time and they were just going to go and they heard it crackin' and they run and they got to the edge of the cave-in and Dad was the only one caught in it. 1929. Just two days before Christmas. That was a tough Christmas that.

It was dangerous just living near the mine. When James Dunsmuir said there was a water problem at Extension he was correct. There was a severe shortage in summer. A fire at Frank Beban's sawmill up the mountain from the camp spread to the surrounding bush. Soon the ridge behind Extension was ablaze and the fire was creeping down the slope toward the houses where the people, knowing there was insufficient water to fight the fire, watched in horror as it threatened to surround them and ignite the mine powder house.

> They had a good bucket brigade here, but in summertime there wasn't much water. There was a lot of houses burned down but they always seemed to be able to save the ones next door. But when Chinatown burnt, they didn't save any there. Because the houses were touching each other, the whole street went. Must have been twenty or thirty houses all burnt down one day there. In 1930 or '31.

Like all coal mining settlements on Vancouver Island, Extension had a Chinatown where single men – some pithead workers, others working for Beban's sawmill – lived in boarding houses. In the evening Chinese men balanced sticks on their shoulders to carry two coal oil cans full of water from the spring that flowed out of the hillside by the road connecting Chinatown to the mine.

There were black people living in Extension too. Some say they came as strikebreakers in 1913; others say they came much earlier from the United States via Salt Spring Island. References to "Coontown" and "Negroes" are scattered through the stories told about Extension. The one about the black driver who was fooled by a ventriloquist into thinking a mule could talk was a particular favourite despite its racist overtones. The story reflected the attitude of the general population in those times towards unusual people whose numbers were too small to be threatening.

The majority of the populace was white: Italians, Finns, Slavs, Austrians, Hungarians, Danes, but mostly British and many of them interrelated.

> Now the Bowaters came to Northfield in 1885 or '86 and the Greenwells came in about 1884 straight to Nanaimo from Nova Scotia and before that the family came from Newcastle-upon-Tyne. So they opened them mines in Extension and the Bowaters moved there about 1898 and the Greenwells around the turn of the century. They were a family of seven children. And three Bowaters married three Greenwells. And two other Bowater girls married two Moore brothers. It was a common thing in those days.

> We were like one family up there. Was some outsiders come in but not too many. And most of the time if you sit alone in the beer parlour in the Tunnel Hotel, maybe two or three guys at the next table, they would tell you, "What's the matter? Fence around? You know they want everybody to join in one big table. 'Specially a man named Bowater. He wants everybody together.

They had a dance here at the Knights of Pythias Hall almost every Saturday night. They had special trains on and people used to come up from Ladysmith to the dance. And vice versa, they'd have dances in Ladysmith and then special trains for Extension people to go down there to the dance. In the old Agricultural Hall there.

The interdependence of the two towns persisted as long as there was coal to be mined. James Dunsmuir had sold out to Canadian Collieries (Dunsmuir) Ltd. in 1910 – some said because of the thirty-two deaths in 1909. The operations of the company extended to include Number Four mine, south and east of the main pithead, and Number Six to the north. Number Eight at McKay Lake farther south and the New Vancouver slope opened in 1926, the latter to extract pillars and top coal from Old Number One. There was lots of coal and it looked as if life would continue as it had for a long time.

Then in 1928, Number Eight closed, followed the next year by the New Vancouver slope. In that same year the company closed the entire operation, reopened it and then shut it down permanently on April 10, 1931. There was no shortage of coal, just a shortage of markets. Small groups of miners continued to work the deposits. Chamber's Mine worked all through the Great Depression on one side of Old Number One, and Beban Mine worked the other. The company had just skimmed off the cream, and there was lots left for those who wanted to risk mining it and trying to sell it.

We put in about seven or eight years doin' that. Goin' in from the outside, from the surface in, we could get near these pillars because we had the plans and could see where they were. We only went in about three hundred feet I guess because we were encountering blackdamp. See, this here blackdamp would come seepin' out of the old workins'. Of course you can't breathe that. We had a little fan run by a car engine.

We didn't make any money. We were all partners in it, so if one didn't make any money, the whole works of us didn't make nothin'.

I was workin' with a bunch when we drove through right into this old mine. I looked up and I could see light. This was probably the old air shaft from the first Extension mine. Anyhow, there was a pick and shovel standing there. Musta been there for a long time standin' against the wall. I touched it and it fell apart. Not a breath of air in there in all them years.

After the Extension mines closed, they kept me on to look after things and one of my jobs was going up the steep hill behind the mine to where the fan was at the air shaft. I was supposed to go up there at 12:00 each night, but I was afraid of the dark and cougars and bears and I sure hated going up there. One night I met these two eyes shining out of the forest. I could feel my hair standing up on end. It turned out to be an old mule that they'd turned loose after it wouldn't work any more.

The miners of Extension had been turned loose too. Many left to find other work but some stayed and commuted to other mines. They or their descendants still live in this unique place. Never beautiful in its heyday, it now sits quietly in its small green valley, its inhabitants connected to its past by memory or family lore.

The most stirring time in Extension's history was during the riots that took place in the summer of 1913 during the two-year-long strike that closed all the mines on Vancouver Island. The people of Extension were particularly militant during that strike and in the rebuilding of the union twenty years later. The seeds of that militancy were bred in the discontent engendered by the management policies of the Dunsmuirs and the tragedy of the 1909 explosion and nourished in the unique character of the miners of British extraction who lived in the valley above the mine.

> As long as I've been around Extension, the people there were more militant. I suppose it was the Bowaters and the Greenwells and others who were more militant than average. It would come from their historical background. There were other militant areas in Europe – the Ukraine, Croatia and the Slavic states – but I think myself the most vocal and clear were the ones from the British Isles because of the language. It's pretty hard for a man who speaks the Slavic language to express himself in a country like this. And they all congregated there in Extension. People of a common mind, they join together. And I suppose the Armands were militant people too. The people of Ladysmith were not as militant; they were just "good guys," good supporters.

> The fact that they were so far out in the country – it makes them more militant. They live close together and they get some good discussions going.

In the rough houses amongst the stumps of the trees sacrificed for mine timbers in the shadow of the smoldering rock dump, in the Tunnel Hotel and Magistrini's Store, people talking, listening – good discussions and common grievances can yield an explosive mixture.

> Extension was kind of a rough town, a good place for a good time. You know what I mean? They used to have some good dances and they were very, very friendly people. Very friendly people.

Chapter III

The Black Track Mines

Just over the ridge and a little to the southeast of Extension is another valley and another coal camp. But with the main line of the E & N Railway paralleling the main street, South Wellington is not as isolated as Extension is. Nor was it a one-company town. All of the major Vancouver Island coal companies worked the Douglas seam in this valley – five mines called the Black Track Mines after the coal-strewn trail that led from South Wellington to Nanaimo.

The line of pitheads of the Black Track Mines also runs parallel to the railway and the beginnings of both were intertwined. When Robert Dunsmuir agreed in 1883 to build the E & N, he received a land grant of eighty thousand hectares that included "all coal, coal oil, ores (except silver and gold), stones, clay, marble, slate, mines, minerals, and substances whatsoever thereupon, therein, and thereunder."

The grant contained virtually all the land on the eastern half of Vancouver Island from Victoria to Campbell River except for certain areas already spoken-for, most notably the 2400-hectare HBC grant around Nanaimo and the smaller parcels of land, or Crown grants, that individual settlers had claimed before 1883. Mineral rights came with Crown grants, but families like the Grandams, the Richardsons, the Thatchers, the Emblems, the Becks, and the Bowaters were not interested, at least at first, in mining – they wanted land to farm.

> My grandfather, Sam Fiddick, was a conscripted sailor and the first time he came to Nanaimo, it was on an English warship. Then he came back overland and met my grandmother, Elizabeth Grandam, and got married. She had come here around the Horn. They settled up just off the Nanaimo River on the old homestead there. They had two hundred-acre blocks of land. They were settlers and had settlers' rights so that meant they had coal rights. They didn't know there was coal there when they first took it. They never went that far back in the woods. Coal was on the other side of the E & N track.
>
> There were prospectors all over. You can go anywhere up above South Wellington, above the E & N, and there's prospect holes. The coal was right outside in outcroppings.

Also in 1883, the Nanaimo Coal Company opened the last of the Southfield mines, so named because they tapped coal deposits in the

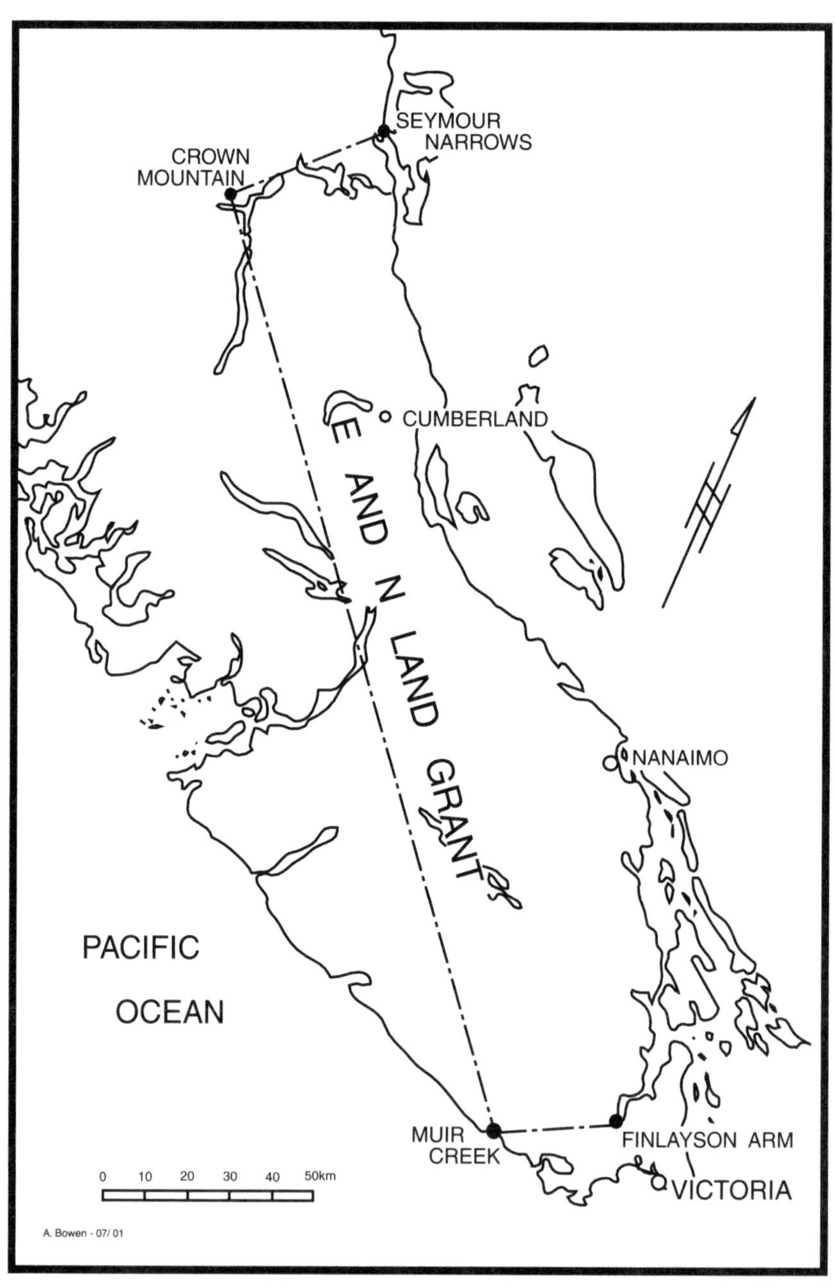

The E & N Land Grant

south field of the company's territory. The opening of the last and most southerly Southfield mine had marked the beginning of mining in the South Wellington valley. Although the Southfield mine yielded good quality coal before it closed in 1902, it was important to the residents of South Wellington only because it filled with water, as did almost all the abandoned mines on central Vancouver Island. A flooded mine becomes an enormous reservoir of stagnant water, a reservoir of destructive power if its walls are pierced.

The Southfield mine having shown the promise of coal farther south, Robert Dunsmuir bought mineral rights in the valley from two settlers, James Beck and Sam Fiddick, and opened the Alexandra Mine in 1884. Although the mine was often mistakenly called Alexander or Alexandria, Dunsmuir had named it in honour of Alexandra, Princess of Wales. By then, of course, the E & N grant had made Dunsmuir owner of all the unsettled land in the valley.

A strike closed the Alexandra mine almost immediately, but it reopened in 1896 and produced coal long enough to take South Wellington into the twentieth century before it closed and filled with water. Shortly afterward, the settlers with Crown grants in the valley found themselves in court trying to prove their ownership of the mineral rights.

> If you took the land up before 1883, you got Crown grants and after 1883 you bought from the E & N, see? And you were issued E & N deeds which reserved everything for Dunsmuir.
>
> Now the way I understand it, if you hadn't finished paying for your place by 1883, then they wanted to claim your coal rights.
>
> It was gonna cost a little bit of money and Charlie Fiddick and a bunch more of them had got fed up and then Ralph Smith, the Liberal that wasn't in but had been, he said, "Boys, I know you're entitled to this but you can't beat money. You'll just be throwing away good money."
>
> They had to go to court over it and lost several times. And the poor devils didn't have any money to be goin' to court all the time. And there was a man by the name of Jim Hawthornthwaite. He was a socialist Member of the Legislature. So there's a bunch of them got together and he says, "I'll take the case to the Privy Council."
>
> Well the outcome of it was, it went to the Privy Council and they won. In England. The settlers won. But most of them never got anything out of it because they didn't sell it. It's still there.

The Fiddicks were one of the families that fought the Dunsmuirs all the way to the Privy Council. Sam had sold his rights to Dunsmuir years before, but his wife, Elizabeth, had a Crown grant too. Their farm and the Richardson family farm next door lay between the Southfield and

Alexandra mines. In 1907 the Fiddicks began to mine their coal under the company name South Wellington Mines Ltd.

> Fiddicks was sellin' the coal just out of wagons. They were deliverin' some and then people from Nanaimo used to go by this Black Track here and get the coal from them as well. Then the Pacific Coast Coal Company bought them out and they bought the Richardson property too so they had two slopes going down into each mine but ending up at the same tipple.
>
> Mrs. Fiddick was a good businesswoman. The Richardsons sold out for spot cash but Mrs. Fiddick, she had her head screwed on right and she said, "No, I won't sell out. I'll take a cent a ton for every ton that comes out of that mine and anything over two feet of coal must be worked." And she had a man named Boyd that used to come around there every month to see that there was no places left. And she made a pile of money. Her son, he used to go there and check the weight pretty near every day.

Pacific Coast Coal Mining (PCCM) named their two-slope mine the South Wellington Colliery. To tap an outcrop of the Douglas seam that began west of the E & N and dipped to the east, it was necessary to tunnel under the railway. It was also necessary to build a separate twelve-kilometre-long company railway that ran south from the pithead on the west side of the E & N, past the coal camp, then under the E & N, and east to the company washing, storage, and shipping facilities at Boat Harbour on Stuart Channel in the Gulf Islands.

At the pithead offices, just over a kilometre north of South Wellington, a store, a supply house, and a large stable housing thirty animals shared a fenced-in area with fifteen company houses, each nine metres square and each so identical to its neighbour that without numbers the tenants had difficulty finding their own homes. The houses on Wall Street, as it was called, rented for $3 or $4 a month.

There was work for everyone in 1910 at the big new mine. As the slopes advanced farther and farther underground, and the levels ate into the coal, the need for mine timbers caused the same decimation of the local timber that had occurred in Extension just over the ridge.

> Timber 130, 140 feet high was all logged down to the sawmill by Beck's Lake, all dragged down there. Dunsmuir owned much of the timber and they had men cuttin' and splittin'. Lots of farmers cut it off their own property too. Later on Frank Beban had the contract for the mine timbers but he had Chinamen doin' the work. The white man didn't get too much work in the loggin' business.

The white man was busy digging coal in the South Wellington Colliery. And so were fourteen-year-old boys.

> It took me half a day to get used to being in the mine and after that it's just like being anywhere else. You get used to it. You got into the old PCCM by

walkin' down a slope. The men'd go so far and sit down because we had safety lamps then see, no electric light, and you had to sit down until your eyes became accustomed to the darkness.

Illuminating the dark tunnels of a mine was only one reason to have a lamp. The other reason was to detect gas. Because all coal mines have explosive gases that are dangerous if exposed to an open flame or even a spark, it is necessary to know when gas is present.

I can remember seeing the oil lamps. You know they had the little oil lamps they put on their hat. My father'd come home from the mine and his lamp had fish oil in it and it smelled to high heaven.

We used to use dogfish oil in our lights. Little pitlamps. With a cotton wick. And we used to pay 35 cents a gallon for the oil. We used a gallon a month. And we used to have a little cadger in our back pocket, see, to fill the lamp every now and again. Yep, that's the light we used in the mine.

My mother lived on that little island in Departure Bay. Her dad had the contract to supply the mines with fish oil. They used it in their lamps and they used it in the machines. Dogfish oil for the lamps and ratfish oil for light machinery.

Many a times there was little holes in the rib – that's the side of the wall. There'd be a hole in there and if there was any water in there and there was any gas, you could hear it fuzzin', fizzin', brrr, brrr, dzzz, dzzz. And many a time I used to put my light in there and light it just to see it burn. There wasn't enough to blow. The gas has got to be mixed up with a bit of air before it'll blow.

Because an open flame in a coal mine is very dangerous, other types of lamps came into use. The carbide lamp produced safer light by dripping water on calcium carbonate to produce acetylene gas, which the miner ignited *inside* a glass-fronted chamber with a flint. The most commonly used source of safe light, however, was the safety lamp. On Vancouver Island, the one of choice was the Wolf lamp. But the safety lamp had as many disadvantages and advantages.

I've got an old safety lamp here. A match'd give a better light. Man's best friend. Inside there they have three gauzes. This lamp, no matter how explosive the gas is, blackdamp or carbon-dioxide or anything, this lamp won't ignite it. It was locked before you went into the mine with a magnet. It couldn't be opened down in the mine.

A first everybody had these. Winch kids had to pack two, a spare one in case somebody's light wouldn't work. The miners carried them on their belt or in their hand. My father carried it in his teeth all the time. Eight pounds. If your light went out, you're runnin' around the mine in the dark. You couldn't see nothin'. Well you knew right and left and you got your foot on a rail and followed it to where you wanted to go.

> First thing, you were searched. You could take chewin' tobacco or snuff but no smokin' tobacco or matches. And the boss checks your lamp to see if it's locked and sees if it's burning then he'd go and blow all around the lamp and if he could blow that lamp out, it's no good. Take it out. And when you got to where you were working you'd put your lamp close up high there and you see that flame's getting' lower and lower, you get the hell outta there. But an electric light will burn no matter where it's at, so it don't warn you. It'll give you a good light, but it wasn't safe.

> After 1917, we had electric head lamps and the fire boss was the only one that had a safety lamp. For testing for gas. Before day shift he would have been around the mine testing each place to see how it was for gas before we went down. But I seen lots of time when the gas was right down to the floor. Not see it, but I mean to say you could sense it.

The fire boss would be checking for firedamp, a mixture of air and methane that rose to collect in pockets where the roof was uneven. Turning the lamp flame down low, he would hold it up near the roof, and if firedamp was present, a blue halo or cap would appear around the yellow flame. Methane gas usually escaped slowly from within the coal and from above and below the seam, but occasionally it hissed out under pressure.

"Damp" is the term miners apply to any dangerous gas. Blackdamp or chokedamp is the most lethal because it kills by exploding or by asphyxiating. A mixture of carbon dioxide and either air or nitrogen, it leaks off the coal and lies near the floor. It can kill either by exploding when near a naked flame or by making its victim sleepy and then depriving him of oxygen.

> You gotta watch out for the blackdamp. It will explode and kill you and if you're not careful the afterdamp will come and kill you again.

After a mine fire or a firedamp explosion, one of the dangers is afterdamp. Odourless and very dangerous, it is mine air with a high concentration of carbon monoxide. Rescuers looking for injured miners had to wait until the ventilation system cleared the mine of afterdamp.

Proper design and maintenance of the ventilation system was crucial. Each mine had at least two connections to the outside air: the upcast, return air, or fan shaft and the downcast, intake, or main shaft. Fresh air entered the intake shaft – usually the main entrance to the mine – sucked in by a huge fan that sat over a second or fan shaft. The fresh air had to travel in one direction through every tunnel and crosscut and into every working area in order to sweep all types of damp out of the mine. In order to guide the air in to and back out of every nook and cranny, the "brattish" man constructed and extended brattice boards or canvas curtains to direct the flow.

Stoppings were another indispensable part of the ventilation system. Heavy stoppings, constructed of five-centimetre-thick timber, rock or brick, were permanent barriers designed to block off old workings, workings where gas collected. There were also stoppings that swung open where trips of coal or groups of miners had to pass through. The doors could be opened and closed again quickly to allow the flow of air to resume its proper path. Some mines employed young boys called trappers whose sole job was to sit in a cubbyhole in the wall and open and close the stopping. It was also possible to short circuit the flow of air by opening a stopping to redirect fresh air in time of an emergency.

But there were emergencies that had nothing to do with explosive gas. Vancouver Island mines were vulnerable to flooding when they were close to abandoned workings. The danger increased when there was a flooded mine on either side.

In 1913, when Pacific Coast Collieries (PCC) bought the PCCM mine, the manager, Joe Foy, applied to the Ministry of Mines in Victoria for maps of the two flooded mines that bracketed his own. Western Fuel Company, successor to the Nanaimo Coal Company and owner of the Southfield mine, obliged; Canadian Collieries (Dunsmuir) Ltd., successor to the Dunsmuirs and owner of the Alexandra mine, refused. When the Southfield plans arrived, it appeared as if there was a thick layer of rock between it and the PCC mine. But Mr. Foy didn't notice that the scale of the Southfield map was different from the scale used in mapping the PCC mine.

On February 9, 1915, one hundred miners were on the job. They had known for several days that they were approaching the border of the company's holdings, but they believed they still had fifty-five metres to go. And they believed that a further 125 metres separated them from the actual workings of the old Southfield mine.

But at 11:30 a.m., on Number Three level off Number One slope, a miner's shot blasted through into the old mine.

> Well there's no pressure until there's a release, see. It's stagnant water there with no weight. But once it gets an opening then it's got force. Old Southfield was above the PCC mine and you got a downward pour through this hole that the blast had made. It made a hole in the solid rock floor you could bury a chair in and it took out rails and wrapped them around like bedsprings.

> Well, just how shall I say it? The water trapped seventeen men who were lower than where the water came in. There was much grief at that particular time. Men was rushed from all sides, pumps was rushed in from wherever they could get them and they started pumping the water immediately. But there was no hope of saving the men that was trapped. They was trapped and

drowned. There was some that got out with the water right up to their necks. They just barely got out of the mine.

William Gibson, who survived the Extension mine disaster of six years before, went to work – and died.

The water rose eight metres on the vertical or 152 metres on the slope. Manager Foy raced to the mine when he heard what had happened. Thomas Watson, working his first day, followed Foy into the mine to help the manager warn miners to get out as fast as they could go. Then a sudden gush of water already carrying several coal cars thundered up from below swallowed them and swept their bodies to a higher level.

There was another case of heroism too. Another man named Bill Anderson got out and he said to the men sitting there, he says, "Is Bob Miller out?" He was his brother-in-law. And they said, "No, Bob is not here." And so Bill went back to see if he could find or see Bob, possibly help him and he too was trapped and died in the mine. Just a straight case of heroism.

Some men were able to struggle through the blackness and the rising water and make it to safety, but the cries of the men less fortunate than they were reverberated in their ears and would for a long time to come.

From all over the district, miners, wives and mothers, children, and doctors converged on the pithead, each one hoping the men below had found a safe place on a ledge or pile of rock above the level of the water. While pumps laboured to lower the level of the water, pulmotors waited to resucitate the drowning. Then, more than four hours after the first rupture, rescue attempts ended when water burst through from a second hole.

Twenty men drowned on February 9th, but the twenty bodies did not come to the surface in sacks until April 30th. It had taken over two months of pumping to recover the bodies.

I had to go and identify my uncle's body when they brought it out. He stayed with us all the time 'cause he never married and if it hadn't been for my mother's patchwork on his overalls, I would never have been able to identify him. The bodies was all swollen up, no hair on, eyebrows was gone after being so long submerged in the water. They had an awful job bringing them out. They had to wear gas masks and everything. It was warm weather, that's what made it worse.

They found a guy sittin' there with his arm around a post and the other arm around his son. And he was sittin' up on the brushin'. Oh yes, they were both dead. Drowned. Never let go of his son.

Most of the fellows that were drowned were the fellows that formed the orchestra that we used to use when we were on strike. See, it happened soon after the Big Strike. And I remember quite a few of them. Before the strike was settled there was a dance or something like that every night in the South Wellington hall. They were good musicians, good singers, mostly Scotchmen, one or two Welshmen.

After the inspector of mines examined the plans of the two mines, he blamed the discrepancy in the map scale in his 1915 report. The people of South Wellington had other theories.

It was greed you see. They talked about maps and surveyors and things but that was a good piece of coal and the company was after it.

They should have known there was something wrong because my uncle had been off sick for a week, said he couldn't eat his lunch for the odour and stink of the place and the miners thought it was stagnant water that was causing this odour. You'd have thought then that they would have started to drill ahead. They called it swamp water. You could smell it.

For two weeks before the flood, miners had been vomiting as they approached the face. They had reported this to their union president. The *Free Press* opined that it would be hard for the United Mine Workers of America (UMWA), weakened by the loss of the Big Strike the year before, to get witnesses because they risked being fired and blacklisted. "The miners' union," wrote the editor, "while still organized, is not as powerful as it formerly was, so that President Foster is somewhat limited in his ability to take action."

The long weeks of waiting for the water to disappear prolonged the agony of the survivors. Children asked for their fathers; women going about their days quietly would suddenly burst into tears. One boy remembered the flood because his mother refused to make him a birthday cake. When the waiting for bodies ended, the funerals began. Fraternal orders like the Foresters and the Eagles conducted funerals according to lodge ritual for members who had contributed monthly to a burial fund; leftover money from the fund went to the widows to cover immediate expenses. The company gave a lump sum of $1,500 to the dependents – $75 per dead man. Surviving miners passed the hat, but long-term survival was up to the widows.

There was nothing for the widows except they go out and scrub floors or take in washing or take in boarders, which many of them did. You could sue the company if you had the money to do it.

My dad was killed in 1922. There was Mothers' Allowance by that time and for our family that amounted to $55 per month. We had a fairly good house in the standards of those days with hardwood floors and so forth but [the next year] the house burned down. It was insured for $3000. So my mother was cut off from the Mothers' Allowance because of the $3000. They let her spend $400 to buy another house and the balance had to be used up at $55 a month until it was all gone and then she got the pension again. So this is how she lived. She did a lot of washing and a lot of sewing, minded families that were sick. She was quite resourceful. And naturally she did all our clothes too. We learned to hunt and fish. We shot deer, grouse, pigeons, you name it. Picked a lot of berries in the woods.

It was the house next door that got on fire and our house was close to it. We were able to get most of our furniture out of it but there was no fire protection. All you depended on was what you could carry with a bucket and most of the water was from wells. This was in July and the wells ran dry in the summer because the mine underneath took all the water away from the wells.

The mine may have taken the water away but it gave the town its sustenance. South Wellington was in its prime in the years during and after the First World War. Most of the almost three hundred inhabitants depended on the PCC mine for their livelihood. And many of them were militant union supporters. In the aftermath of the Big Strike, Western Fuel and Canadian Collieries had refused to hire union men. But the PCC mine, knowing that they were experienced miners and likely to work hard, welcomed them. So families like the Greenwells and the Gilmours had moved to South Wellington in the fall of 1914.

The camp had several distinct areas. In the main town there were houses, boarding houses, hotels, a bank, a barbershop, a shoemaker's, a post office, and a dance hall. Just over a kilometre north of the main town, near the PCC tipple, was Wall Street, the company-house district that had become notorious during the strike as the famous bullpen where the strikebreakers lived. And then there was Scotchtown or "Mushtown", named in honour of all the porridge consumed by its inhabitants.

Well that was across the lake. They sold property off the Beck estate over there. A lot of it sold in five- and ten-acre farms. They called it Scotchtown because of Jock White who was a pit boss. And one Scotchman would tell his friends to come to Canada and go and see Jock and he'd say, "D'ye come frae Glasgae?" "Yes." "Come out in the morning." And that's how it got the name of Scotchtown. They fetched a whole lot of Scotchmen out when a Scotchman was boss.

In the early days, this whole area was 90 to 95 percent British. The Scotchman and the Englishman they come out here and took the land off the poor Indian. But you ain't got much pure blood left now. Look at me. I'm half Scottish, quarter English and a quarter Welsh. I'm married to a Finn girl. I got half Finn kids and part English and part Scotch and they get married again back into somethin' else.

There were some Italians but there were more Balkan people. South Wellington got a lot of Austrians and Yugoslavs. The Swede immigrant went to the logging camp more.

Now the Finns came out and they congregated. They weren't too far from bein' like Doukhobors. They wanted to clutter up into their own thing and make their own laws and one bunch went up to Malcolm Island, Sointula. There was a lot of 'em in Chase River, you know, they had their own hall and then they had their own hall in Ladysmith. Some of them weren't satisfied with their own country and they weren't satisfied with what was here. But they were a great bunch for having a leader. And this leader of theirs took a bunch of them to Russia and one of my wife's brothers went. And when they got to Russia they started to tell the Russians how to run the country so they put 'em up against the wall and shot 'em. It was in the early twenties years and years after the Sointula business.

There was Finns and Italians, Belgians and Scots. They had their own lodges and their own social groups and those things played a big part in the lives of those people. Those things have their value in that they give a feeling of cohesiveness, a feeling of belonging and that's good. But unfortunately, too many of them didn't reach out to the rest of the community so that they could learn English well. You know, some of the older people, the mums particularly, weren't very good in English.

They came from the Old Country with the idea of making a better life for themselves. Probably they moved out of sheer necessity and certainly with the idea that out here they could improve things and that their children wouldn't have to undergo what they had gone through.

It didn't seem to matter where they came from, a lot of 'em liked to drink. Many of our local celebrities whether they're English, Scotch, Finnish or what, on a weekend were quite capable of standing toe-to-toe with anyone else.

By golly there was always a bunch at the hotel. Well, them days beer was cheap. And the guys'd go to the hotel and fool around and get drunk. Well, South Wellington, when it was running full blast, it was quite a place.

On Saturday night people liked to have fun. If the mines were closed and things were tight financially, they would stay home or take their children with them to dance at the local hall for 50 cents each. The bootlegger at the Bucket of Blood would sell cheap liquor to anyone in need of his wares.

But when everyone was getting a paycheque, the place to go on the Saturday night after payday was Nanaimo, ten kilometres to the north.

We had three trains each day, all the way from Ladysmith or wherever you came from, stopped in South Wellington, Stark's Crossing or you walked or took a horse and wagon, bicycle and then when things got more opened up well they got trucks and buses – jitneys we used to call 'em.

The bus went to Nanaimo at 5:00 p.m. and came back at 10:00 or 11:00. Saturday night in Nanaimo was a must for some South Wellington miners. They would check in on the dance at the Oddfellows Hall, have a few beers at the beer parlour and shoot the breeze.

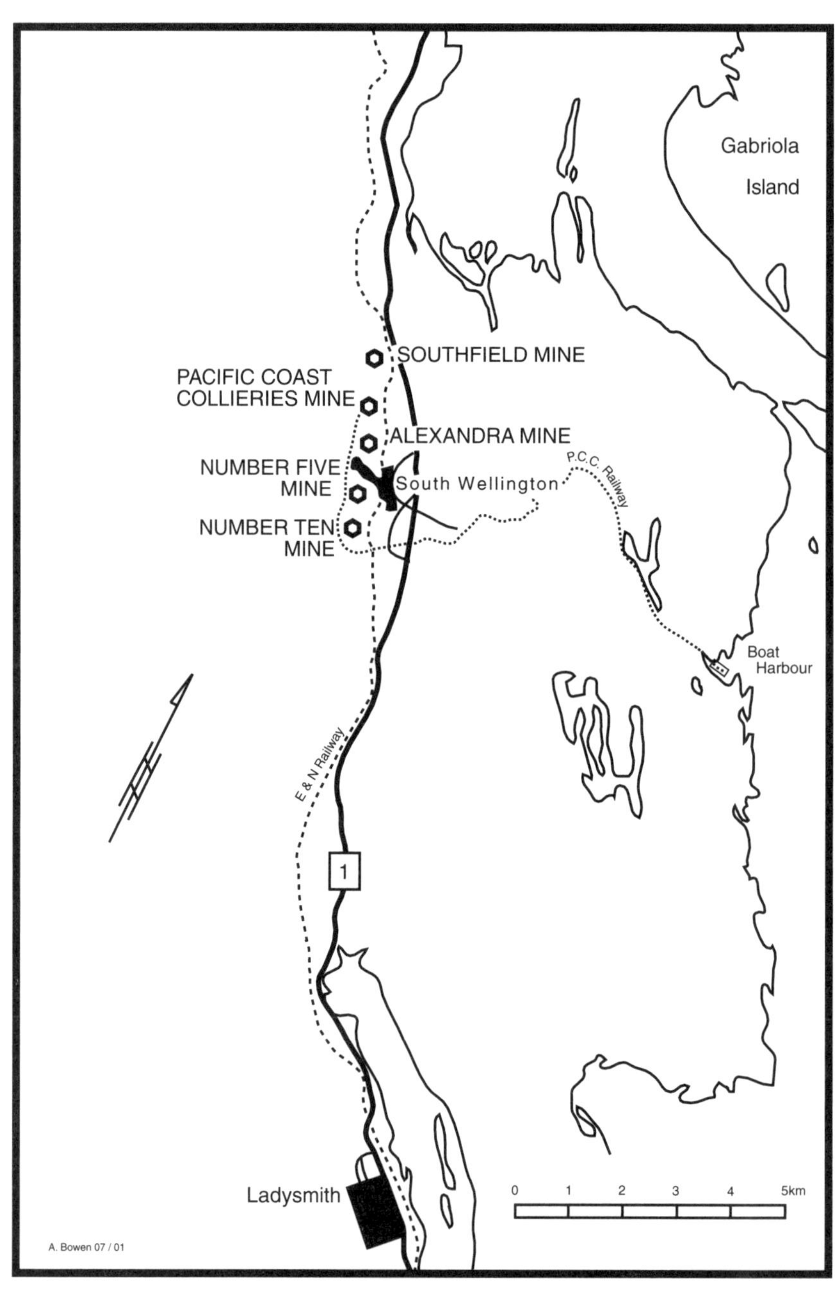

The Black Track Mines

Most of the time, the residents of South Wellington stayed in their valley, living in relative isolation and with no indoor plumbing or running water. To wash clothes and bathe children and coal-black husbands, it was necessary for women to pack water in from outside and heat it on the stove.

> In the summer time when the wells went dry, we used to have to carry the clothes down to the water and wash them there and then carry our water for around the house from the creeks, maybe a mile away, Chinese style, a stick over the shoulder and a five-gallon pail on either end. Smaller kids carried smaller pails and would help carry the clothes to the creek and so on.

Wells went dry in the summer, mines dried up and houses and hotels burned for lack of water to quench the fire. But in the winter and spring there was too much water.

> We used to come home from the mine wet through from water comin' out of the roof. They used to give us 25 cents a day for that. When you got home, you took off your clothes and you used to 'ave a line across the kitchen to 'ang them up to dry.

When the pumps had cleared the PCC mine of the floodwaters in 1915, the miners went back to work. In 1916, however, they reached the boundaries and began to work back, pulling pillars. One year later, the mine closed, the pumps stopped and the waters flowed back in.

But Canadian Collieries was already planning to open their Number Five mine right in the middle of South Wellington. Like other Black Track mines, Number five had a slope that would take the men down into the mine. This was high seam coal and there was lots of it – well drillers for a nearby hotel had gone through almost four metres of good quality coal. The mining method of choice for Number Five would be pillar and stall.

From the very first tunnel dug in any coal mine until the very last, the roof and walls have to be supported. Until recently that job was done with timbers. As each miner advanced the coal face in his work place, he had to take time away from digging the coal – the only source of his income – to support the roof and walls with stout timbers. He took two, notched each at one end and leaned them against the rib or wall on either side of the tunnel. Then he notched a third log or stringer at both ends and placed it on top of the other two to form an arch, being careful to tip the upright timbers slightly inward to make the arch more able to absorb the pressure exerted by the mine. He repeated this process every metre or so and laid lagging or boards on top of the stringers, extending the lagging beyond the last stringer to give his head some protection from falling coal or rock in the untimbered work area.

A new mine usually began operations with the latest refinements in equipment, especially at the tipple. Number Five had a revolving dump to empty coal cars, shaker screens with two decks and facilities to carry boiler coal to fuel the power plant. Four boilers produced steam to drive the hoist engine, provide electric power, and run the compressors for drilling.

Chinese workmen and young boys worked at the tipple. Hard-fought agreements from the past kept the Chinese from working underground in most mines, and provincial law was supposed to keep the young boys above ground until they were sixteen. But most parents, mine officials and inspectors were flexible when it came to boys going underground. While the official age for starting work below the surface had risen sharply from the days of child labour in British mines in the early nineteenth century, and had continued to creep up as parents realized the benefits to their sons of staying in school longer, there were many miners who worked the coal mines on twentieth-century Vancouver Island who were twelve years old when they got their first mining job.

> Well when you were a boy that was all that was around here them days. I don't even think there was a mill around here. That's all we had to look forward to.
>
> They couldn't keep me out of the mines. I was bred a coal miner. My father was eight years old when his father was killed and he was diggin' coal when he was eleven. He was workin' on the face and he worked in the mines all his life. I left school of my own accord to work in the mines. They wanted me to keep going to school but I said, "No dice."
>
> The only reason I went down the mines is because my dad joined the army in 1917. I was fifteen years old then and like father like son, I guess. I should have continued my education but in those days when you lived in the country you had to pay to go to high school. Anyway I had a lot of chums that was workin' in the mines and my mother needed the money. The mines was the only thing there was in those days.
>
> I never had my own money. Whatever I made, I gave to my mother. Whenever I wanted a dollar, a couple of dollars, ten dollars, whatever I wanted, to buy a suit or pants, all I had to do was go and ask her and she'd give me what I wanted. No questions asked, but I never had my own money.
>
> The thing I remember most about the pithead, we got $1.71 a day. We worked harder than the mules. Lifting heavy timber and dumping rock. They paid according to your age, not according to what the job was worth. And there was great advantage taken of young people because they really worked them.
>
> I started picking rock and I can remember my first wages. It was about $2.81 per day, not hour. I was raised on a small farm so I guess maybe I was fairly strong and large for my age. And I was only there for about three or four months when I got promoted to another job. So I got an increase in pay. It would be over $3 anyway. I stayed on that job until it got too small for me

> and then I kept naggin' at my father and he finally took me down the mine, coal mining at the face with him when I was about seventeen. I didn't have a miner's certificate but at that time you could go and work with your father as a sort of helper or backhand.

Like apprentices everywhere, the new young miners were the butt of jokes and pranks, some of them as old as mining itself. Most of the new boys took this in their stride and went on to dish out the same treatment to the next batch of novices.

> The older men pulled little tricks on you. Tell you to go out for a broom to sweep up the tracks and to go look for Jim Crow. A jim crow is a tool to bend track but you didn't know that so you'd be looking all over the place for him.

> We were on a place which is pretty low and I was just a green kid and one of the old miners comes up to me and says, "Come on and we'll give you a good place that's high so you won't have to work on your knees." Little did I know that the special place was a "nigger head." That's a place in the roof where a piece has come off and left a hole. A little pocket of gas would sit up in there. In about five minutes my legs started to cave and I went down. I didn't know what was wrong with me. The other men started laughing. It was a joke.

The first underground job for a new boy was often driving a horse or mule on haulage – the most frustrating job in the mine for a driver unable to deal with the animals' eccentricities. But if he could work with them, he might find himself with a job for life, thus preventing him from eventually becoming a digger, the job that made the most money.

Expert drivers preferred mules to horses. In a low place a horse would shy or bolt. In a tight place a horse would try to crash through. Mules could be difficult, but horses were impossible.

> It was a horse I had the worst time with. There was a little dip and he kept slowing down so he couldn't get the car up the other side. I'm behind the car and he's trying to shake hands with me every time I'm trying to get past. He's kickin' and tryin' to bite. So I says, "I'll fix you." I give him a few taps on the backbone with this sprag. A sprag is what you put in a wheel to stop it. Well he started to lean on me. Tried to imprison me against the rib. I gave him a few kicks and he kept tryin' to hold me there. So I took a long sprag and I sharpened one end to a point. Then when he started to lean on me I just put it up against his ribs and let him lean. Well that battle went on for days between him and me and one day I was so bloomin' mad I took a drill that the miners use and I put it up against his belly, gave it a couple of turns and he got up, poop, poop and away he want. After that, every time he'd stop I just rattle the tool and away he'd go.

> Sometimes I used to get mad at them horses. I killed one of them in Number Five mine. Well, I had to do all the work. I'd practically be pushing the cars and his tail chain would be just hangin' slack, not pullin'. So I worked like that for a few days and then I says, "That's enough." So I clubbed him across

the ears and he went down. I went outside and I says to the boss, "Mr. Martin," I says, "my horse passed away."

Horses and mules were valuable to the company. Mules in particular were expensive because the company had to import them from Kentucky and Missouri. The company protected its investment by providing comfortable, clean stables at the pithead or shaft bottom. Western Fuel had a large farm at the base of Mount Benson behind Nanaimo where the company grew hay for feed and where injured animals convalesced. When mules worked a double shift they were always fed and watered in between. When a man worked two shifts back-to-back, he had to depend upon his fellow miners to give him food and drink.

A mule was worth more than a man. Oh absolutely. In them days when a miner got killed, they'd just throw him on top of a car. The missus had to come and get him.

Rats were another form of animal life in the South Wellington mines, but the miners were glad to see them because they believed that rats could sense danger. If the rats left, it was time for the men to leave too.

Always called a rat a man's best friend in the mine because they're there to warn you. I seen it.

But no one warned the miners of what the 1920s would bring. Even though Vancouver Island mines reached their peak of production in this decade, oil was having a noticeable effect on the markets for fuel. Number Five South Wellington closed down in 1924 for that reason. The closure lasted for twenty months. Then work resumed and continued for the next five years uneventfully with only the advent of a motor ambulance in 1926 to provide a small measure of excitement.

"Pinches" and "wants" – frequent occurrences – were an Island-wide phenomenon. A beautiful two-metre coal seam could pinch down to almost nothing in no time. When this happened, the company had either to pay the miners for prospecting in the hope of re-establishing contact with the seam or to abandon that tunnel, pull out the pillars as the miners worked back and look elsewhere for coal. In 1929, in Number Five, they chose the latter. Considerable development began in the direction of the old Alexandra mine that lay just north of Number Five. In 1930, miners carefully bored a hole in the barrier pillar between the two mines and all the water that had filled the old mine since it flooded in 1901 drained into a large sump formed by the lower workings of Number Five. Pumps took the water to the surface.

And when the pumps got all the water out, it were standing the same as if the miners had just left it. It was unbelievable. You see, the water preserves everything. It preserves the timber so that the air can't get at it. I admired

> the miners because the walls were so straight you could have papered the wall. They dug it all with picks and when they were putting up timber, they just notched it a sliver. It was just beautiful.

Crews cleaned up and regraded the main slope, relaid the track, installed an electric hoist; a narrow gauge steam locomotive began to carry Alexandra coal to the Number Five tipple a kilometre away. New roadways yielded up an excellent though friable coal. Twenty-nine years of retirement had not improved the water problem, however. The roadways were very wet, even muddy, making working conditions unpleasant. The only good thing about a wet mine is that the water keeps the level of coal dust and thus the danger of explosions very low.

But in 1932 and 1933, the miners began to encounter outbreaks of spontaneous combustion. The affected area had to be sealed off. Then miners encountered another pinched seam. This time nine hundred metres of rock tunneling failed to relocate the missing coal. Miners began pillar extraction in earnest. In 1934, two more outbreaks of spontaneous combustion forced the separation of the two mines once more and their abandonment the following year.

But fifty years of continuous mining had provided the inhabitants of South Wellington with relatively steady employment close to home and they were not about to quit mining coal just yet. Since 1927, the Fiddick family had been running a small but successful operation on their side of the old PCC mine. Surface cave-ins into the old workings had made it obvious that the coal lay just below ground level. The grade of the new slope was easy enough for a horse to haul up the coal cars. The cave-ins provided natural ventilation. Year after year until May 1936, a workforce of three or four extracted a steady amount of coal to be sold locally. For another year in 1938-39, three men working on a co-operative basis scratched yet more coal from that original claim, acquired unknowingly when Elizabeth Grandam married Sam Fiddick and they settled down to farm their land.

The Richardson property on the other half of the old PCC mine, saw similar activity. Six men reopened the slope in 1928, and a converted automobile engine powered the hoist until 1933, when the three Richardson brothers abandoned the old slope and divided their father's property between them. One of the brothers, Bill, and his wife, Dolly, farmed their land and started a new mine that Bill named after his mother, Ida Clara. The new slope had a small tipple, a quality of coal that happened to be in demand at that time and the only female miner for miles around.

> The Ida Clara was a good mine. It brought out some real good coal, you know, the real grade stuff. It had a real steep incline and we had a car engine

to pull up the little coal cars. We had a siding off the E & N and whenever we needed a big coal car we just got in touch with the E & N and they would drop one off for us. The mining railroad was eighteen gauge, very small, and used old cars from the Ladysmith smelter. There was a lot of water. And we'd have to go down lots of times and use a barrel and buckets to bail it out. It wasn't millions or thousands of tons, you know, it was maybe ninety-eight a month or two hundred. Something like that.

It was the Great Depression and Dolly and Bill were doing everything they could to make ends meet. They dug coal, milked cows and put up hay, sometimes by moonlight.

My mother used to say to me, "I never seen a woman that would do something like that." And I said, "Well, I enjoy it." And when I was cutting wood well I've seen me split two cord a day. You get real hardened to it and you can just cut wood like cake.

The Fiddick family had a remarkable woman too. The same Elizabeth Fiddick that drove such a hard bargain with the PCCM mine for coal royalties was able to set aside a comfortable amount of money, some of which she used to help people during the Big Strike. Legend has it that she also took on Mr. Dunsmuir.

One of her cows got killed on the railroad. So she stopped Dunsmuir's buggy on the road and she says, "I want fifty dollars for that cow, Mr. Dunsmuir." "Well," he says, "It's not my fault your cow got killed." She says, "You're supposed to fence your railway." Then she walked right between the wheels of that buggy and says, "I'm not movin' out of here 'til you give me the fifty dollars in my pocket." She didn't take him to court but she got her fifty dollars.

It was people like the Fiddicks and the Richardsons who made this area unique. First settled as farmland, the ground beneath South Wellington yielded up large quantities of coal for each of the major coal companies and for many small operations as well. It was here that ordinary people took on the Dunsmuirs over settlers' rights, and just as Mrs. Fiddick bested Mr. Dunsmuir, so they eventually won.

The Black Track mines included Number Ten mine. Its story belongs to a later time, but it was in every way typical of these mines: All were entered by slopes, all were mined by pillar and stall, and all were very wet.

I went back to Southfield with another miner and we explored it using chalk to find our way back. We walked in a long way, looked up and saw a light up above. Here was a room all bricked in with a tall chimney, a car, and a grate. Someone told me it was what they used to ventilate the mine. We found a coat hanging on a nail. We touched it and it fell to powder. We found a pick and shovel. We touched the shovel blade; it disintegrated, but the wooden handle was all right. The timbers were beautifully made, so straight and smooth, three feet apart exactly. Built like a picture frame.

Chapter IV

The Town That Fought Back

There is a Dunsmuir Street in Vancouver and one in Nanaimo. There are Dunsmuir castles in Victoria. But nowhere does the name have more significance than in Cumberland. There, far removed from the Nanaimo and Wellington coalfields, clumped along the main avenue that bears the Dunsmuir name, sits a town whose entire existence used to depend upon the Dunsmuirs, father and son. It was they who transformed a tiny coal camp into an important coal-producing community and it was against their antiquated labour policies that the Cumberland miners fought. Although Robert Dunsmuir and Sons sold out to Canadian Collieries (Dunsmuir) Ltd. in 1910, many former miners believe it was the Dunsmuir legacy of neglectful and dictatorial management that forged the militant labour force that led the long fight for union recognition on Vancouver Island.

The wide and muddy expanse of Dunsmuir Avenue was the focal point of a coal-producing area that included Cumberland, Bevan, and Union Bay. Cumberland had begun life as Union Camp, named after the Union Coal Company, a consortium of Vancouver Islanders including at least seven miners. Unable to muster the huge amount of capital necessary to mine the coal and transport it to tidewater for shipping, the partners had sold their shares to Robert Dunsmuir in 1883.

At the time of his death in 1889, Dunsmuir had only begun to develop the Comox seam. So it fell to his son, James, to oversee the transformation of Union Camp into Cumberland, a boomtown with all the excitement and problems that the arrival of hundreds of newcomers, mostly men, can bring. Cumberland in its heyday was a place of many bars, muddy streets, and a few adventurous children.

> In the old days, the streets of Cumberland were nothing but mud. It was awful. Then they built wooden sidewalks and we liked those. We went up and down them sidewalks fifty times a day just to try them out. There were sort of little creeks underneath them in them days. But the mud. . . One man had a horse and wagon to deliver groceries and he'd come over to our house for lunch and my brother and I had to go out to get the muck and that off the wheels because they wouldn't turn around. Oh you never saw anything like it.

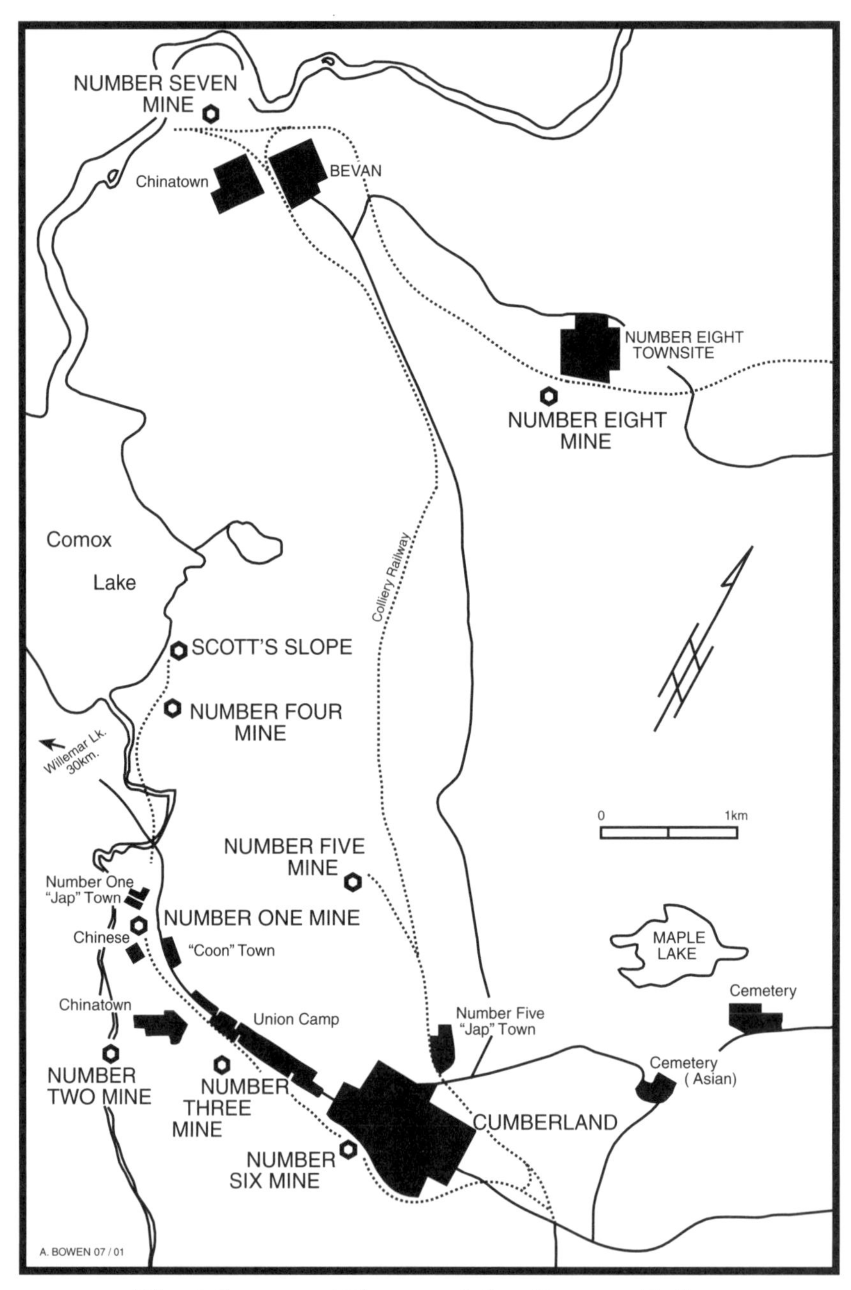

The Mines and Towns of the Comox Valley

The mud did not deter the drinkers from getting to the hotel bars. There had been several hotels to choose from when the camp sprang up overnight in the 1890s, through Prohibition in the 1920s, when the bars sold only "temperance" beer or bootleg whiskey, and up to 1933 when fire destroyed one whole side of Dunsmuir Avenue, including the Union Hotel. In modern times, even though the mines have been closed for over a generation and the Victory and Vendome hotels are gone too, the Waverley, the Cumberland, and the King George are ready to serve their customers every day just as they were when the law required them to close on Sundays.

> I sold beer in the King George Hotel on Sundays. When a Mountie was coming to town he had to notify the city policeman. So as soon as the city policeman got to hear that the Mounties were coming, he'd phone the hotel and say, "Don't sell anything today."
>
> The one you had to watch for was the Union Hotel because that was outside the city and the RCMP used to come down and look in the door so we'd keep the door locked. But the Waverley and them, they'd keep the door wide open, city police, our own police, we'd can the bugger if he did anything. He was scared.

Even at its peak population of thirteen thousand, Cumberland was exclusively a coal mining town. Families lived in the company houses that lined both sides of the avenue in the original Union camp, their well-kept gardens a testimony to the hard-working, practical people who lived inside. Or families rented accommodations in the newer part of town: two rooms above a store for $8 a month or company houses for $12 a month when the average miners' wage was $3.30 per day. Bachelors lived in cabins at nearby Comox Lake or boarded in town.

> In them days there were four or five boarding houses in Cumberland. The hotels were boardin' houses too. There was quite a few beer parlours and they all kept boarders. Board was $25 a month; room was $4.
>
> The miners' entertainment? Drinking beer and cribbage. They had a picture show there but for the miners it was mostly drinking beer and cribbage. We played crib for beer all day Saturday and Sunday. Even after I was fire boss I would still play crib. They'd go out to Comox Lake fishing and a lot of them went hunting during the season. There was a bowling green and a library and a billiard room and a dance hall. So it wasn't always drinking but we did spend a lot of time at the hotel.

Prostitutes plied their trade in the hotels and at Halfway House, a notorious establishment conveniently located midway between Cumberland and her sister town of Courtenay. There's a legend that says irate wives from Cumberland marched on the place and burned it to the ground.

On Saturday night, there was always Chinatown – a separate community of three thousand souls, almost all men, imported by the company to dig coal. Seen from the mine camp, it was a densely packed collection of one-and two-storey wooden shacks. Crooked streets snaked between tailor and herbalist shops, grocery stores, restaurants, laundries, theatres, fan-tan and lottery houses, judo clubs, opium dens, and privately owned houses where groups of men pooled their resources to save money on room and board.

The citizens of Chinatown conducted their lives as close to traditional ways as possible. Sustained by strong family ties in China, these men were working in Canada to send money home to wives and children left in the care of large extended families. Because their original intention was to save enough money to buy land or a secure retirement when they returned to China to stay, they adapted to Canadian ways only enough to survive. Secret societies helped newcomers to adjust. The Dart Coon Club, its building the tallest in the camp, provided for its members' physical needs and gave moral support in the face of unemployment or discrimination.

The Chinese at Cumberland worked in an emotionally charged environment. They were the only men of their race allowed to work below the surface, all the other Chinese miners on Vancouver Island having been excluded from underground work. In the heated days after two explosions – one in Nanaimo in 1887 and the other in Wellington in 1888 – took over two hundred lives, seventy-five of them Chinese, white miners wrung an agreement from the two companies involved. The Nanaimo Coal Company and Robert Dunsmuir and Sons reluctantly agreed to the white miners' demand that Chinese be excluded from underground work. When the Cumberland mines opened a few months later, however, Chinese miners went underground to dig coal, albeit for lower wages and usually in separate working places.

This did not mean that the Dunsmuirs were enlightened employers. In this, as in all their dealings, business or otherwise, they took the course that gave them the largest return on their money. Labour was often scarce on Vancouver Island and white labour was becoming increasingly demanding. Chinese miners worked hard and worked cheaply. The manager of the Cumberland mines found them to be as safe as any other miners, careful, faithful, and obedient. Considering where his paycheque came from, the manager's assessment may have been suspect, but it was a fact that Number Two slope, which employed Chinese miners exclusively, had the lowest accident rate of any coal mine in the province. Any white man's view of the Chinese worker had to be based on his own particular biases, however, because no one seemed interested in really understanding the Chinese worker or determining how capable and safety conscious he really was.

There had been Chinese working in the Nanaimo coalfields as far back as 1867, but it was Robert Dunsmuir who first employed them in large numbers in his Wellington mines. As soon as he purchased the Comox coalfields, he pursued a policy of hiring Chinese miners. They came either directly from China or through local Chinese merchants acting as agents, some of those hired having worked on the building of the Canadian Pacific Railway (CPR) and the E & N.

The miners argued that the Chinese made the mines unsafe and the politicians added this to their anti-Asian arguments. From 1887 to 1896, year after year, one or another stalwart member of the legislature rose to propose an amendment to the Coal Mines Regulations (CMR) Act which would exclude the Chinese from all underground workings in all mines in B.C. They argued that a miner who did not speak or read English could not read the regulations posted in the mines and was therefore a danger to mine safety, ignoring the fact that there were many non-English-speaking immigrants from Europe working in the mines unopposed. They could not read posted regulations either. Sometimes the amendments to the CMR passed, but they were always disallowed by higher courts and were always ignored by the Dunsmuirs anyway.

The actual mining of coal is a precise business. The Dunsmuirs could import all the Chinese labour they wanted, but this did not make the Chinese into miners. Ironically, it was white miners, the very people whose jobs were threatened, who had taught the Chinese how to mine coal.

> The white men hired Chinese themselves to work for them. See, a few of them had contracts and were paid by the amount of coal they dug. They had the Chinese work for them and paid them very low wages. So all this time the Chinese were learning all about mining by watching the white miners and the company got wise and hired them themselves.

> The old-timers would take maybe two or three working places and they were allowed to hire their own Chinamen. If he had three places, he would hire six Chinamen. He told them where to drill the holes and all he had to do was blast. He would shoot the loose coal and then he would say, "All right Jim, you go ahead and load up." It was the white man exploiting the Chinese. That was in the early days in Cumberland. Oh they had it soft. He'd maybe pay them a dollar a day. There was quite a lot of them doing that. Too many of them.

Practices varied with the mine owner, but the situation with regard to Chinese in the mines at the turn of the century was as follows. There were no Chinese working underground in the Nanaimo, South Wellington, and Extension coalfields. Chinese worked at the pithead, at the harbour coal-loading facilities, and in the sawmills supplying mine timbers. They lived in Chinese enclaves at Ladysmith, Extension, and Nanaimo. In Cumberland, Chinese miners worked in all aspects of the digging and transporting of coal and lived in the largest Chinatown in North America after San Francisco.

The Dunsmuirs, having ignored the government when it tried to pass laws preventing Chinese miners from working underground, used another government law to ensure the loyalty of their Chinese work force – the infamous head tax.

The head tax was a political weapon used to reduce the number of Chinese immigrants to B.C. The majority of voters in the province in the late nineteenth century viewed the Chinese as a faceless horde that would overcome B.C. in wave after invasive wave from a never-emptying reservoir of humanity across the Pacific Ocean. They also represented a more immediate threat in their willingness to work for wages lower than most white men could live on. It was therefore a popular move when the government imposed a tax of $50 on each Chinese immigrant wishing to enter the country. By 1904 this had risen to $500, a sum equal to about two years' wages. Then a 1924 law stopped Chinese immigration altogether, thus preventing wives and children from joining men already in Canada.

> There's not that many Chinese women folks then because the Chinese had to pay a head tax. So unless you are a merchant, you can't have your family here, you see, so that's why the majority of Chinese here were bachelors. Not real bachelors but might just as well have been. They got their family back in the old country and that's a long way away.

The head tax influenced the character of Chinese life in B.C. in two ways. First, it made sure that the overwhelming majority of Chinese resident in the province were single men and as such did not establish homes or adopt Western customs. Second, it gave the Dunsmuirs a way of binding the Chinese labour force to them in Cumberland, forcing them to work as strikebreakers. The company extracted 50 to 75 cents a day from a wage of $1.25 towards repayment of a man's head tax. As the amount of the tax increased, so did the length of time it took to pay it off. The Chinese were hardly able to defy an order to work when half their wages went to repayment of the head tax and the other half had to be divided between living expenses and putting aside money to send home. In addition, unlike the white miners, the Chinese received free water, light, and oil and free miners' lamps, a fact that increased their obligation to the company.

With the goal of eventually returning permanently to China, most Chinese kept their customs intact and adapted to Canadian culture only where necessary. And it was the more exotic aspects of their culture that intrigued their white neighbours. Chinese New Year, in particular, drew many interested onlookers.

> That's when they would dress up in costumes. They were beautiful, magnificent, just like in Vancouver. Satin with mandarin collars and long sleeves. You wouldn't know they were the same people. They'd set off fireworks and put lights up. A shipment used to come in here once a year from China. That's probably why there was a custom's office in Cumberland, to clear the stuff for Chinatown. Mostly foodstuffs, rice and they'd bring in their own vinegar and salted eggs and all the Chinese herbs, plus preserves, ginger.

> Us white kids always loved Chinese New Year. They had marvellous fireworks and they used to give us a kind of candy thing. It was like a nut but it was soft inside and very palatable.

> On the last day of the year, just before mealtime, you go and see anyone who owes you money. You ask them for it. After that you don't dare to go and ask them because that's taboo. The first day of the holiday, that's New Year, then from that day on up to the tenth day, everyone is on holiday if they can take the time. Then they have another celebration, a big feast and that's the time for them to go back to work again.

Some white miners were more than interested onlookers when it came to the fan-tan and lottery houses in Chinatown. Guards stood watch so the white miners could hide behind sliding panels if the police came. When it came to using opium, however, outsiders were more likely to speculate than participate.

> My brother and mother used to get mad at me for going down there because the Chinese smoked opium. They wouldn't give it to white people. That was for their own use.

> They had those silver opium pipes, water pipes. Oh, about two feet high and it bubbles, you know. They'd sit there and dream about China. What a life they had.

In 1908, William Lyon Mackenzie King, federal Minister of Labour and future prime minister, discovered accidentally that not only were the Chinese buying and smoking opium openly, but two firms were paying a $500 yearly licence fee to manufacture it legally. Advertisements for opium shared space on the front page of the *Victoria Colonist* with rice, tea, and sugar. King reported his discovery on June 26, 1908, and the House of Commons wasted no time. On July 13th Parliament passed a law prohibiting the import, manufacture, and sale of opium for other than medical purposes. It seemed to have minimal effect, however. The port of

Union Bay, outlet for all Cumberland coal, was practically a free port for the entrance of both Chinese labour and opium. Nanaimo, Ladysmith and Boat Harbour were almost as bad. As late as 1921 the Chinese in Cumberland were still smoking opium.

The Chinese always remained a mystery to the white people of Cumberland even though the two communities lived side by side for forty years. A medical doctor who grew up in Cumberland's Chinatown characterized the relationship in this way, "We were individually praised but collectively hated." The praise usually took the form of admiration for their skill and for the logic they applied to the amount of work they did for their meagre pay.

> They were good workers but they only done so much. They didn't work as hard as white men. The Japanese did but not the Chinese. They figured they got low wages so they did just so much. At first they were supposed to work in places that the white man wouldn't work but eventually they worked anywhere.

> I worked in Cumberland Number Four mine and there were Chinese diggers. All Chinamen on the longwall. And they were good diggers too. I was on contract, so much a car. Take a car and get an empty, take a car and get an empty. So one morning I went there and the Chinese weren't loading. A gang would have one bossy man, maybe one that could talk a little better English than the others or something. So I went to the bossy man and I says, "What's a malla, Jim." We called all of 'em Jim. "What's a malla, you no loadum car?" "Oh," he said, "you sabbee before him dollar half. No him dollar quarter." The company had cut two bits off the wages. So the Chinese decided not to fill their shovels as full each time and slowed down.

> You didn't really have any trouble speaking to them but some of them were pretty cute. You didn't speak like you do ordinarily. You tried to speak to them so they'd understand you. They would tell you they didn't savvy, but they savvied all right.

> If they laid off one Chinaman, the next day there wouldn't be a bloody Chinaman around. Five hundred Chinamen. They wouldn't go back to work 'til that guy got his job back.

> I worked with the Chinamen and I know that they were about the best men I ever worked with. I think they were the best mule drivers I've ever seen. I never seen them beat a mule. This one Chinaman drove the same mule for two years and if you drove that mule after the Chinaman did you couldn't do nothing with him. After he left, a white guy got this mule and Christ, the bloody mule pretty near kicked his eye out, pretty near kicked the shit out of him. So he knocked the mule's eye out and beat him black and blue. After about a year, year and a half, that Chinaman came back lookin' to kill that white man.

> The "other" race, they were well spoken of, well respected by the other miners. And I think it was mutual. I can remember Sam. I've forgotten the old Chinese name. But I was working on the pithead and when he came by, he said, "New boy?" and I said, "Yes." "What's your name?" I told him. Are you Davie's son? Ah good man." He had to go by me to get fresh water from my grandmother's well to take to the office cooler. From that day on, I always had fresh water.

But underneath the good will and sometimes grudging admiration was an undercurrent of racism in the mines and in the towns, most notably in the use of the word "Chinaman."

> We were scared I'm telling you in case we got too involved with them. In our own family we were not allowed to use the word "Chink" in referring to the Chinese. We must say "Chinaman" or "Chinese." That sort of phraseology was not acceptable. But lots of people used it.

> They were the most friendliest people you ever wished to meet. When we were children going to school the vicar once came to us and said, "Oh, you shouldn't play with them. They will be mean to you." I said, "No way, they were so friendly."

As anti-Asian sentiment in the province reached its zenith in the 1920s, it found expression in Cumberland after an explosion in 1922. A rumour that a Chinese miner lit a match that caused the explosion led to the exclusion of Chinese miners from the Cumberland mines.

> When a new seam started in Number Five in 1932, the Chinese and Japanese weren't allowed in that mine. When Number Four finished that finished them. They weren't allowed in any new mines. Government rules. The only Chinese in Number Five was one who worked in the barns where the old seam was.

> The government that stopped the Chinese from coming in here was the Conservative. There's lots of Chinese still have a grudge against that. To me, I don't. It's a time when prejudice against the Chinese was high then, so therefore, the Chinese was the whipping boy. You see what I mean? That was back in 1924. But now, things change.

As mine after mine forbade them entry, the Chinese began to drift away. Some went only as far as Union Bay to work arranging the coal as it slid down chutes into the holds of deepsea freighters. Others ended their days in the ghettos of Vancouver and Victoria. Many of the buildings destroyed in a 1936 fire in Cumberland's Chinatown were already deserted. The Chinese had visualized a different leave-taking: either they would return to China with money for the future, a feat few accomplished, or their bones would come home seven years after their death.

Many Chinese died in Cumberland in the major disasters of 1901, 1903, 1922, and 1923 and in isolated mishaps over the years. Some of the most memorable spectacles seen on Dunsmuir Avenue were the funeral processions from Chinatown. Led by the town band hired for the occasion, wagons decorated with streamers punctured with holes to allow bad spirits to escape carried dead miners through Cumberland to the cemetery outside of town. Mourners placed the miner's lunch pail beside his body in the grave and, when the dirt filled the hole, left large amounts of food to feed his spirit. After the body had been in the ground for seven years, the living dug the bones up, polished and boxed them, and sent them to China, where they would rest in earthen jars in the side of a hill in a macabre reunion beside the bones of his family.

Three thousand Chinese once lived in the crowded, ramshackle enclave outside Cumberland. They were admired for their ability to work hard, but resented for their acceptance of lower wages, and misunderstood as only a group of strangers who remain apart and seem to be indistinguishable from one another can be.

> I can remember the Chinese. I can remember my grandma sitting on the porch down in the camp watching the Chinese and Japs go by...counting, and she would maybe count 300 and she would say, "I'm not counting anymore." One after another.

To many people there was no difference between the Chinese and the Japanese. The miners, however, detected a difference in attitude.

> If you tried to ask a Chinaman any questions, "I no savvy" and that's all you got, but the Japs would talk.

> I think the Japanese were more ambitious than the Chinese in a way. The Chinese were more tight-mouthed, but the Japanese were more advanced in a way.

The Japanese and Chinese cemeteries outside Cumberland are near to each other but much different in appearance. Buried on a beautiful treed hillside are many of the large number of Japanese who died in the mines in similar numbers to the Chinese, but in the Japanese graveyard there are women and children too. The Japanese came to North America to set up homes and raise families. They came to stay.

Families meant children and children needed schools. In Cumberland a school for the Japanese children was at a point equidistant from the two Japanese communities named for the mines they were nearest: Number One Jap town and Number Five Jap town. An active program of exchange with the white school enriched the children of both races. But still, the Japanese remained as separate as the Chinese although white people admired the beautiful round pagoda in Number One town where Sunday holidayers on their way to Comox Lake could buy rich homemade ice cream.

It was the ice cream and a "darn good baseball team" that Cumberland missed when the federal government deported Japanese families to the interior of the province after Japan attacked Pearl Harbor during the Second World War. Now Number One town has disappeared and only a few buildings from Number Five remain behind the scrub on the north side of Cumberland.

> At the time the war broke out that was when they cleared the Japs out. Yeah. The Japs were good people too.

> When my mother and my two brothers and I went walking after Sunday school, we went to Number One Jap town for ice cream. We'd go from the mine camp, past Chinatown, Coontown and down the hill to Jap town.

A dozen black families brought in, some say, to be strikebreakers in 1912, lived in Coontown. Others say they had been miners in the United States and had emigrated to escape racial harassment, ending up on Salt Spring Island. Whatever their story, they remain an exotic oddity in Cumberland's story, disappearing as anonymously as they came, leaving only one of their number, John Brown, to give evidence that they had been there.

The modern ear recoils from such terms as Chink, Jap, Nigger, and Coon. But the speakers used the terms unselfconsciously, the offensive words reflecting a lack of sophistication rather than deliberate racism. The ambivalence and uncertainty show in this remark made by one old-timer.

> People weren't prejudiced against all those people. It made no difference. They were friendly as can be. We had no trouble whatsoever with the niggers or I would say the dark people and the Chinese and the Japs. In no way. In Cumberland we were all friendly.

From its very beginning until the day the last mine closed, Cumberland and its sister community of Union Bay were a one-company operation. The vast Dunsmuir resources enabled the two towns and the railroad connecting them to develop at great speed. Coal from Cumberland came by company train to Union Bay to be washed and stored in bunkers then loaded on ships bound for San Francisco, San Pedro, and San Diego principally, but also for Oregon, Washington, Alaska, Petropavlovsk, and Hawaii.

Union Bay also had a coking plant. To construct the ovens necessary to produce coke from coal dust, a brick works in Victoria combined fire clay from the Cumberland mines with ordinary clay. In 1899, the Union Bay Colliery began to produce its own bricks, drying them in two kilns housed in a two-storey building. Since the closure of the Union Bay facilities and the dismantling of the coke ovens, the bricks have appeared in a number of homes and most notably in the façade of the liquor store in

Cumberland, where the distinctive gold-coloured, wedge-shaped bricks remind people of the days when the mines were operating.

The first mines were Number One and Two slopes and Number Three tunnel, all situated on Cumberland's southern perimeter. Never very successful mines, the three closed at the beginning of the twentieth century in favour of Number Four slope, which had opened near Comox Lake in 1888, Number Five mine, which had opened in 1895 north of town, and Number Six shaft, which had opened in 1898 on the southern edge of the town. Number Five and Number Six were always meant to be interdependent, the 186-metre-deep shaft of Number Six would provide the return airway for Number Five mine, and the shaft of Number Five would supply the intake airway for Number Six. As soon as the tunnels of the two mines connected, the ventilation system for both mines would work properly. But the connection was not completed until 1908, seven years too late to help sixty-four miners who died as a result of a burning brattice curtain on February 15, 1901.

The ventilation fan drew the heavy smoke from the fire throughout the mine, suffocating twenty white men, thirty-five Chinese, and nine Japanese before the fire was extinguished. The inquest could not determine the cause of the fire, management was deemed blameless, and a new rule forbade more than nineteen men to be down either mine until they were connected.

Such a ruling could not prevent an accident, however. Only the numbers were different the next time, and the inquest was nearly as vague about the identity of the guilty parties. The foundations of Cumberland trembled on July 15, 1903, when gas exploded in the tunnels beneath the town. The inquest determined the cause to be "negligence on the part of the sufferers" who were, in this case, sixteen dead Chinese miners unable to defend themselves against the charge.

All miners in Number Six used safety lamps, which the lamp man locked before they went into the mine, but the inquest found that an explosion would not have occurred even with the large amount of gas, if there had been no other lights present. In the pockets of the dead men their would-be rescuers found matches, tobacco, and cigarettes. This was a technical infringement of the rules, but many miners took cigarettes to work. They rolled their own – eight fit perfectly into a Peppermint Lozenge tin – and took them to work to be smoked before they went down and as soon as they came up. And many miners chewed tobacco in the mine. The presence of smoking equipment in a miner's pocket was not a foolish thing in itself. Unless he lit a cigarette in the mine. If that was what happened in Number Six, then the combination of a large release of explosive gas, the foolhardiness of a miner lighting a match in a mine known to very gassy, and the lack of progress in digging the connecting tunnel, which would have made proper ventilation possible, caused the explosion that day.

Not everyone believed the story about matches being found or about there being only locked safety lamps for that matter. Number Six had a bad reputation, one that preceded the explosion.

> Some people blame the Chinese for that explosion. Said they were smokin' or something. I heard that there was supposed to have been one found with a match on him. His shoes even blew off in the explosion so how could they find a match? Down there you never find anything.

> My father told me it was so bad in Number Six with firedamp and black-damp. He used the old teapot and fish oil lamp and you had to go into the place packin' your lamp just midway on your body. If you put it down it went out in the damp and if you put it up well you had a fire goin'. Flashfires. His brother was cautious. He met him comin' home from work and he says, "I'm finished. I'm not going down there again. The tools are down there if you want to go down and get them." They brought their own tools then. So Father went down and got them of course and he quit too and they went to Number Four and about two weeks after that old Number Six blew.

Carbon monoxide killed four men in Number Six on June 3, 1917. Twelve days later the company stopped work in the mine due to a shortage of labour. It was wartime and hard to find able-bodied men to dig coal when so many were in the armed services, but this shortage may have been caused by the fact that miners sometimes reach the point when they refuse to risk their lives in a particularly unsafe mine. As a miner who left the gassy mine just before the 1903 explosion said, "I'm finished. I'm not going down there again."

Number Six shaft remained open to serve as an airway for Number Five and an avenue for pumping out water until 1931. Then the company removed the equipment and the buildings. A person who knows just where to look can see the capped shaft behind the city garage; 186 metres below is a labyrinth of tunnels that spread out under Cumberland and once caused it to shake on its foundations.

Like a Siamese twin, Number Five had depended on Number Six for its existence. And like its twin, it was gassy. The seams produced methane gas even when undisturbed.

> The mines I seen the most gas in was in Cumberland. I worked in Number Five there one year. You could hear the gas workin' in the coal. Zzz. Zzzz. And the coal gas would be pushin' the coal out. And it's just like a sponge and you squeeze it tight and you see the water comin' out of it? It was just like that. All of a sudden you see the coal burstin' out like that.

Number Five was dangerous. A lot of gas in it. We had to sit there before we got in as far as we were working. Sometimes we'd sit there for an hour or two. Too much gas.

The area at the bottom of the shaft with its stables for mules and horses was the only place in Number Five where the miners could stand upright. The seam was low, usually no more than a metre high. For a narrow seam like this, longwall is the method of mining. Longwall – where the men spend their whole shift unable to stand upright, working on their knees.

Except for one level all the other workings in Number Five were longwall. Some places you got five feet but most of the places were three feet and when a wall would cave you'd have to crawl in and brace it and it would be eighteen inches high. You were lyin' on your belly.

I was a haulage man but I went down and a lot of places these men worked in was two feet, three feet high. On your knees all day and then when the wall would cave, it caved down to about eighteen inches and they'd have to break through and start another cut going to get the height back and they'd be on their bellies. Loading coal on their bellies. And I'd be small and I'd get in there and help them you know. It was tough.

One of the biggest troubles, you had to have special knee pads and they were made of rubber and they sweated your knees pretty bad all the time.

You get used to working on your knees. After a while you don't want to stand up any more. I saw a man shoveling brush and rock into a car and the place was twelve feet high and he'd be down on his knees still shoveling into the car.

There's two advantages to workin' under a low roof. First, you don't have to bend and straighten your back all the time and second, your head's close to the roof so you can hear the slightest unusual sound or movement.

You wouldn't catch me workin' the longwall. Crawlin' around on your belly all the time. That's for snakes.

Because rock encased the seam and because rock removal was expensive and time consuming, companies used the longwall or panwall technique in narrow seams to reduce the rock work as much as possible. Longwall required a deep horizontal cut either at the base of the coal seam or half way up it. At first the miners used hand picks, crawling in and cutting the opening out lying on their sides.

Longwall is like working under a low table and going in the long way. Before they had the machines it was done by hand. The undercut was done with sharp puncher picks and then hand picks and when they get the undercut they drill holes in there and blast it out.

Two or three guys got hurt you know. One old miner had a hand pick drove right up through his nose here. Stuck on the roof and couldn't get him out. I was the smallest there so I had to get the saw and I went in there and cut the pick off. Got him out. Yeah, he survived.

By 1905, there was a machine for longwall that resembled a large power saw blade. The machine cut two metres into the seam leaving a 17-centimetre-wide space. The miners drilled holes for the shots. A poorly set shot produced nothing more than a big bang. A properly placed shot made a dull thud and loosened the coal in just the right way. Loaders on their knees using shovels with straight, short handles loaded the coal onto trough-like pans, which shook to move the coal along to the end, where it dropped onto a conveyor and from there into coal cars.

Pretty hard to describe these coal-cutting machines. They cut like a big chain saw and they're goin' by compressed air. And they had a big chain goin' around with picks on it.

I was one of the diggers on those panwalls. One guy would drill it and the fire boss would shoot her. There would be nine loaders in one longwall and they would all be loading the coal onto pans. One man could load twenty tons a day. The pan was right there by the face so you'd have to work on the other side of it. At first what you had to do was shovel underneath 'til you got a space for the blade. Didn't leave you much room. Sometimes that was pretty tight and noisy. Those pans bangin' all the time.

This shaker shakes the coal down into the car and there's a chunker at the bottom there and he trims up each load until he's got a trip of maybe eight or ten cars. Then the haulers take it away with mules. In Number Five there were anywhere from ten to sixteen men on each longwall crew. It was all rough work.

They took out enough rock to get a car in. That's what they call brushin'. After the coal had been taken out, they blasted the rock to get enough height to get a car in. The car had to be down lower than the coal.

Muckers and timbermen played an important role on the longwall crew. The muckers worked right after the machine crew to clear away rock and debris into the gob or waste area that was already cleared of coal. The timbermen placed supports every metre or so to support the roof as the machine worked up the wall. Timbermen dreaded the phenomenon called "first break."

First break occurred on a newly opened longwall when the roof became too large to support itself. The weaker layers broke and crumbled around the timbers and then the thicker layers of sandstone did the same. Timbers creaked and eventually split, sandstone rumbled and cracked, and the process of crushing and grinding continued until the roof met the floor.

I've seen one. It's really scary especially for fellas that don't know the ropes. You look at the roof and you think, "It'll never come down, it'll never come down. You're just workin' and you hear it crackin' and crickin' and then it stops and then the floor starts comin' up to meet the roof. Well you get the hell out of there. The roof's comin' down but the floor's comin' up too. The posts and that are pushin' down so hard on the side. So instead of lettin' that happen, I says to this fella, I says, "You go to the panwall and ask for a pick. I'll fix it. If it's gonna come down, it's gonna come down." I looked for the loose end and made it come down and it missed us and the other fellas had enough room to crawl through to tell them guys on the panwall to come to the other side.

Funny, in the mines if it's gonna cave it's always after midnight. Between midnight and 6:00 a.m. Always. Must be atmospheric pressure or something.

Aw, baloney. If it's gonna come down, it's gonna come down. It's got nothing to do with after midnight.

Number Five mine had an unusually irregular seam even for Vancouver Island. The longwall method becomes very difficult when a seam suddenly disappears in an upthrow or a downthrow.

Gas added to the frustration. The fact that Number Five experienced no major explosions was more a case of good luck than good planning, at least in the early years. By the 1930s, however, anti-gas measures were routine and strictly adhered to: lime rock dust and water spray controlled coal dust on all roadways; water sprinklers did the same on all conveyors; enlargement of airways improved the movement of fresh air. But it was still necessary at times in Number Five to prohibit all blasting and to inspect it frequently. One year the provincial inspector checked it eighty times.

As the only mine in the Cumberland area working between 1931 and 1937, Number Five had supported many families during the Great Depression, but by 1947, the production face of the mine was so far from the shaft that the cost of maintaining the elaborate precautions was prohibitive. Second World War subsidies, which had kept the mine in production, had ended with the peace treaties. All that is left today is the slack heap and some concrete footings.

The largest Cumberland mine was Number Four slope at Comox Lake, an energetic walk away down a road that passed Chinatown, then Coontown across the road, then Number One Jap town with its ice cream pagoda. Then the road crossed Coal Creek and swung north to come near Number Four slope before reaching the lake with its bachelor cabins and vacation areas where many children passed happy summer days on the shores and in the water of Comox Lake.

But the mine near the lake was not a happy place although many people made a living digging its coal in the forty-seven years it operated. Number Four was a difficult mine, gassy, and prone to flooding and fires.

> I can't tell you any funny stories about that mine because there was nothin' funny about it.
>
> A lot of people in Nanaimo would come up to Number Four and look for a job and they'd see us coming out of there so wet and they'd go back to Nanaimo again.
>
> I was up to my waist in water lots of times. And we didn't get extra for that. One place was on fire and another place was flooded. They wouldn't work under those conditions today. It was burning for years. They shut down the mine and flooded that and thought they put it out but they didn't. They went in there again and it broke out again.

In Number Two slope Number Four mine, fire and flooding followed one another in unending succession. A fire in 1901 made it necessary to flood a section of the mine. The flooding caused caving of the friable roof and walls, which impeded the pumping-out process. Once the water was removed, the wet gob or waste material heated and ignited spontaneously. That meant reflooding with the resulting damage to roof and walls. The alternate flooding and pumping continued sporadically until 1914, when more efficient pumps removed the water before it could do its damage.

It was the fire clay from Number Four that went into the bricks at Union Bay, but it also made the roof friable and caused 60 percent of the accidents. The fire clay came down with the coal, making cave-ins more likely and making it very difficult to keep the coal clean.

Number Four was the last mine to employ Chinese underground. This meant that the mine continued to function at almost full capacity during the Big Strike from 1912 to 1914, but it also meant that Number Four killed a large number of Chinese. Two events occurring within six month of each other caused the greatest loss of life.

The inspector of mines in his 1921 report stated that the workings of Number Four mine had spread out so far underground that the ventilation system no longer worked properly. The inspector also noted the presence of some coal dust and some gas. Number Two slope in particular, notorious for spontaneous combustion, was like a bomb waiting to detonate. Yet when there was an explosion there in the very next year, the inquest absolved the management of the mine of all blame.

The explosion occurred on August 30, 1922 at 3:00 p.m. The residents of Cumberland heard the bang and streamed from their houses down the road toward the mine by the lake. The roof had caved in several places over three kilometres from the portal of the mine, burying men and equipment. Stretcher and ambulance crews rushed to the mine; helmeted Draegermen swung into action, climbing aboard a man trip assembled hastily. The chain of coal cars waiting to carry the rescuers had to go deep into the mine, and no one knew if there'd be another explosion. A doctor climbed on board and was followed unexpectedly by a nurse. The doctor was horrified. "The mine is not place for a woman." Nurse Belotti was adamant. "I'm going with you," she said, "I know when I'm needed."

In the east level pump house 2.5 kilometres in, the rescuers set up a first aid station. Scrambling through the debris, they found men buried to their necks in coal and rock and broken timbers. Some had their legs pinned; some had their backs broken; some were dead, burned as black as the coal they had dug. Four white men and twelve Asians injured—two crippled for life. Eighteen dead: six Japanese, nine Chinese, three white.

> My father was killed in the explosion of 1922 in Number Four mine. After Dad left for work that day, my mother happened to look out the little window over the sink and she could see him turn back. She was Irish and superstitious and she hollered, "Stop Jack, stop. Whatever you want I'll get it; don't come any further." He kept coming, so she hollered, "What do you want?" He had forgotten his chewing tobacco and that was the last he spoke to her.
>
> She was hooking a rug around three o'clock. We kids had just got home from school. She heard the noise of the accident, the bang and she looked up and said, "There's your father, he's gone." And that was it. My brother was in the mine and was knocked down two or three times but he got out and he walked home all that way to let my mother know he was all right. She just kept hookin' that rug. What could she do? They wouldn't have let her go. They brought my father home in a coffin. His big dark mustache was burned right off his face.

The inquest found that debris from a brushing shot had broken an electric power cable and the resulting arc had ignited the gas. The explosion traversed a 180-square-metre area. In exonerating Canadian Collieries (Dunsmuir) Ltd., the inquest ignored the overextended ventilation system, which, if properly operated, would have rid the working area of the gas. The inquest, however, did dispel the rumour that the culprit had been a Chinese miner's cigarette.

But another Chinese miner was the scapegoat for a second explosion six months later. In the same mine on an evening when the gas was very bad on Number One slope, Number Two east level, the fire boss cordoned off the area and sent the crews elsewhere to work. As the inquest report later said, "It is presumed that one Chinese returned to the area, lighted a match for some illicit purpose, and ignited some gas nestling in the rib roof-breaks." This, in turn, travelled in the direction of the intake air current, reached the larger body of gas, and exploded.

That the identity of the "one Chinese" was not specified and the use of the pejorative phrase "illicit purpose" appeared in an official report, shows the temper of the times regarding Chinese in the mines. The true cause of the explosion is still unknown, but the effect of the explosion is: thirty-three white and Asian miners dead.

When the news reached Cumberland, Canadian Collieries rushed a train carrying nurses, doctors and first aid equipment to the site. A series of cave-ins – each one requiring the clearing of debris and replacing of damaged timbers – slowed the rescuers' progress. It had an awful familiarity to it.

Two groups of trapped miners saved themselves by building barricades to block the poisonous atmosphere until rescuers could re-establish the ventilation system. Fifteen Chinese miners owed their lives to James Pinfold who chose a route by observing which way the air was moving and led them safely out of the mine through the maze of tunnels.

> *Nanaimo Free Press*, February 8, 1923
> Cumberland is a house of mourning where women go wailing through the streets for the loss of their loved ones and there is ever a knot of men and women waiting at the undertakers' office to see who is borne in next.

For all that dark winter night and well into the next day, stretcher-bearers brought in the bodies of the thirty-three. Mummy Martinelli, a laughing good-natured Italian man, had died in a ditch, one large arm thrown over the body of a young boy as if to protect him. Martinelli left a wife and children and so did many others. In all, there were twenty-five children orphaned by the explosion. Among the bereaved was the widow Mitchell, who waited for the body of her eldest son, William, the sole support of his mother and five brothers and sisters. When the rescue party reached him, he was seated at the hoist he had been operating, the frayed skin on his hands the only outward sign of the terrible explosion that had sucked the life out of his sixteen-year-old body.

The usual fund to aid widows and orphans received donations, many from other towns; a small provincial government Mothers' Allowance would aid the widows who would manage with a combination of ingenuity and hard work to raise their families. Serving as an example for all was the woman who lost her husband and three sons in the 1903 explosion in Number Six. She raised her remaining four children, all girls, by serving as a midwife, delivering babies at home, and accepting whatever pay the family could spare. A woman whose father died when she was a young had glowing words for her mother's ability to provide.

> There were thirteen children but I don't remember ever going to bed hungry. My mother always cooked extra potatoes and extra meat and at nighttime before we went to bed, she'd take this great frying pan and she'd cut

up all this meat and potatoes and it was the most gorgeous thing you ever tasted. We always had fresh bread and with jam it was better than cake. We used to pick blackberries and Mother made pie and jam. Just thinking about it makes me want it. There was lots of soup all the time. You'd make it from shin beef and when you served it you'd put the soup in a bowl and the beef separate on a plate and with bread and butter and your tea, that's all you needed.

Each time a coal camp suffered a tragedy, it staggered then regained its equilibrium and got on with the business of earning a living. Cumberland went back to digging coal, but the days of the mine by Comox Lake were numbered. Although operations continued for another twelve years and a new slope was even opened, no new development took place in the old workings. Ironically, the miners extracting pillars there were not troubled by methane because all the gas had escaped from the coal years before. As the miners removed the pillars and worked back, the pumps that had struggled to keep Number Two slope free of water were turned off in the abandoned levels.

The opening of Scott's slope near White's Bay in 1930 seemed to promise new life to Number Four mine. Although 1.5 kilometres of narrow gauge railway joined the new slope to the old tipple, there was no underground connection in order to avoid accidental flooding. But Scott's slope was short-lived. The mine employed fewer and fewer men until in January of 1935 a great storm unleashed a large inflow of water. Number Four lost its final battle with flooding water. It no longer paid to pump the old mine out.

A few kilometres north of Cumberland, however, Canadian Collieries had constructed the town of Bevan. In many ways it was a miniature version of Cumberland with its own Japanese enclave, its own Chinatown, and its own mine, Number Seven, close by. Everything in Bevan belonged to the company. Between 1911 and 1912 the company built each one of its 150 houses; the company built its school and the large store and the hotel that accommodated sixty boarders.

The houses were nice enough. A miner and his family could rent a four- to six-room house for $11 or $12 a month. Although the insulation was meagre, there was a fireplace in every bedroom and no shortage of coal. But when the seam started deteriorating in 1918 and miners looked elsewhere for work, the company cut the houses in half and moved them to Cumberland. As long as the renters were prepared to follow their houses wherever they went, they could keep them.

Bevan was a pleasant place to live during its short life. The Chinese and Japanese were more a part of the community there; in fact the Chinese Free Mason Hall was the place where the entire community gathered. All the children went to school together, and the Asian children kept the white children on their toes.

> Oh they were smart. A white kid never beat an Oriental. They had problems speaking English, but they were always on top of the class.

On July 6, 1922, a fire that followed a long, hot dry spell levelled Bevan's Chinatown. Many of the buildings were empty, abandoned when the mine closed the year before. Fifty company houses and the hotel and general store remained, however, waiting to play a part in the revival of the Million Dollar Mystery.

At the same time that Canadian Collieries was building Bevan in 1912, the company was investing large amounts of money just up the road to sink two three-hundred-metre shafts for the new Number Eight mine. The mine was to have the best of machinery designed by British engineers and built in Europe; the machinery was to run with electricity from the new company powerhouse on the Puntledge River close by. The mine was to be a modern one, one that would need less skilled labour and be less at the mercy of unions.

Building continued all during the years of the Big Strike. The plan was for seventy-five houses. In 1914 the mine swung into operation. Then suddenly, on August 5, 1914, the day after Britain declared war on Germany, work stopped. Miners were paid off, the cages and cables were removed, houses were boarded up, and the workings allowed to flood. Save for one man left behind to keep the machinery functional, the place was deserted. Rumour had it that it cost $1 million to build. Since the only explanation ever given for the closure was a falling off of the coal trade, people christened it "the million dollar mystery mine."

> Nobody seems to know why they closed it. Well, they sold shares in the old country in the mine. That had a lot to do with it. And then with the mine being closed, the shares petered out and the people that bought the shares never got a nickel.

The mine lay idle for twenty-two years until, with no more explanation than when it closed, it reopened in November 1936. Twenty-five of the original seventy-five houses were completed; forty houses in Bevan were revived along with the Bevan hotel and store. The people moved in and life began again.

> These were all company houses and we paid $7.50 a month rent, free water and free light up until so many kilowatts or whatever. We were on 25 cycle then. They supplied paper and paint for you. The houses were just shells, there was no insulation or anything. We did have water in the house – just a tap – and that was supplied from a spring. It was a mine prospect they dug or drilled for coal and that's what supplied the water for the whole mine and for this community here.

When we first came we didn't hardly have to go to Cumberland because there was a big store in Bevan and he used to come down here three times a week to see whether anybody wanted groceries. He'd go from house to house. It was the same with the Wilcox Meat Market in Courtenay. They did that three times a week for your meat and they had some vegetables. Of course there was a Chinaman who used to come around with his truck every week with vegetables and fruit. And then people used to come from Denman Island and Hornby and bring apples. I remember there was a man he used to come around selling fish. The grocer from Bevan would get anything in the hardware line and bring it to you. Very few people had cars.

At the peak there were about fifty families. We got our mail from Bevan. There'd be the odd thing for entertainment in the school. We had a two-room school here but our children used to go to Bevan until it closed down. Then there was photo night. A bus used to come from Courtenay if you wanted to go to town on Saturday and on certain nights you could go to the show and they'd draw a name and the winner would get something and they'd have their photo taken.

The mine needed 400 men to run it. Those who could not live at Bevan or Number Eight town, commuted from Cumberland on the company train – a steam locomotive pulling two passenger coaches lined with tongue and groove V-joint and painted a dull barn-red. Two rows of plain wooden benches were good enough for men in dirty pit clothes. The otherwise well-equipped mine had no washhouses until 1943.

The big mine was set to roll in 1937 with all the surface installations in place and the two shafts sunk to the 212-metre level. After the descent in the sixteen-man cage that took less than a minute, the men stepped off into the main entry: five metres wide, two metres high, well-timbered, and lit with electric lights. The stables with floors and walls of concrete had a comfortable stall for each of the fifteen to twenty animals housed there.

Very little mining had gone beyond the main entry area, however. Because the coal seam was only seventy-five to eighty centimetres wide and sometimes as narrow as forty-five centimetres and because it lay very deep under tremendous pressure from the rock above it, the method of mining would again be longwall or panwall. In order to establish the panwall, the first work had to be done by hand.

Number Eight where I was workin', when they brushed the road out, the rock in the roof there, you could see ferns and salal leaves, just as plain as day. Fossils like. You could see the shape of small trees like with bark and that. The leaves would impress right into the rock, you know. It was coal. It would look like a piece of wood. Well this is what coal was at one time, vegetation.

There was shale on the bottom of the coal seam and shale on top. Pretty rough when we started. That was 1937. That was before they had the longwall and it was all pick and shovel and hard work. Just loading into the cars by hand. After they got the longwalls it wasn't so bad. Still hard work shovelling onto the conveyor.

Number Two seam in Number Eight mine consisted of two bands of coal forty-five centimetres thick separated by a band of shale three to twenty-five centimetres thick. In order to remove no more rock than necessary, the working space was no higher than 115 centimetres. Only at the entries to each longwall at intervals of ninety metres was there enough height to stand erect. Small wonder that most miners remember Number Eight for sore knees and stiff legs.

It was another gassy mine, but the industry had learned a lot about dealing with gas by 1937. The key was control of coal dust and superb ventilation. Number Eight's huge Sirocco fan could move 6,400 cubic metres of air per minute, its distinctive sound reassuring to the wives living nearby.

They had a great big fan that was running all the time and if that fan ever stopped they had to get out of the mine because they had to keep the air circulating to keep the gas down.

I guess if something happened at the mine or say the fan stopped all of a sudden or something, the women wondered what happened. Of course, you could always hear the fan going, and as soon as everything would get quiet then you'd know there must be something.

Modern mining techniques seemed to have eliminated the major disasters that had made the Cumberland area a place of such sadness in earlier years. Death still stalked the workings though, looking for a careless move or the slight lapse that would give him a chance to strike.

There was two men killed in Number Eight mine from 1937 to 1953. Well, one was killed with a pile of rock and the other was hit by a coal trip. The first one the fire boss went in and fired a bunch of shots in the coal and you could see the roof was bad and the fire boss told the man, "Now get some timber up there or else pull that rock down." He told the man what to do and he didn't do it. Of course the man went under that rock and it came down and killed him. But the other man was a pipefitter. Now he'd been into the mule barn and as you come out you come into the main level where the coal trips are running. Well he came out there and he didn't stop to look and he got hit with a trip of coal.

Seventeen years was all the time that fate allowed Number Eight mine. On the day in February 1953 when the mine closed, four hundred men were out of a job.

They took everything out, dismantled it, dropped one cage into the shaft and it was terrible because that was a well-organized mine and there was lots of coal there yet. They sold all the stuff for scrap in Victoria.

> When the mine closed down a lot of the houses were torn down and taken away, or moved whole the way they were. A few of us wanted to stay here so they said if we paid for the surveying we could buy them. We got the house, no plumbing but with electricity and a 120 foot by 140 foot lot for $700.

There are a few houses there yet. Some are completely changed. Others look much the way they have always looked: square, functional, a little homely, but secure. The concrete skeleton of the tipple is there too although the massive steel girders are gone. The fan house, roofless and without its fan, sits off to one side by one of the capped shafts. Over by the other shaft are three tall arches rather like the supports of a Roman aqueduct. They lend an air of archeological interest to what remains of the Million Dollar Mystery.

Nanaimo District Museum 03-31

Work weary Extension miners pour off the train in Ladysmith heading for home and a hot bath.

Ray Knight Collection

On the porch of the Abbotsford Hotel in Ladysmith, mourners in lodge regalia attend the coffin of a victim of the 1909 Extension explosion.

Cumberland Museum C160-088

Pithead boys and mule drivers with oil lamps on their caps gather below Extension camp near the mine tunnel entrance.

Cumberland Museum C160-061

Pithead boss Bickerton watches as a carload of hard coal is emptied at the Extension mine pithead c. 1920.

Cumberland Museum C160-076

The picking tables at Number Five South Wellington mine – in winter the open sides of the shed let in the cold wind.

Cumberland Museum C160-077

A pit boss views Number Five South Wellington's contorted seam, his hand-held boss' lamp giving off a brighter light than miners' lamps do.

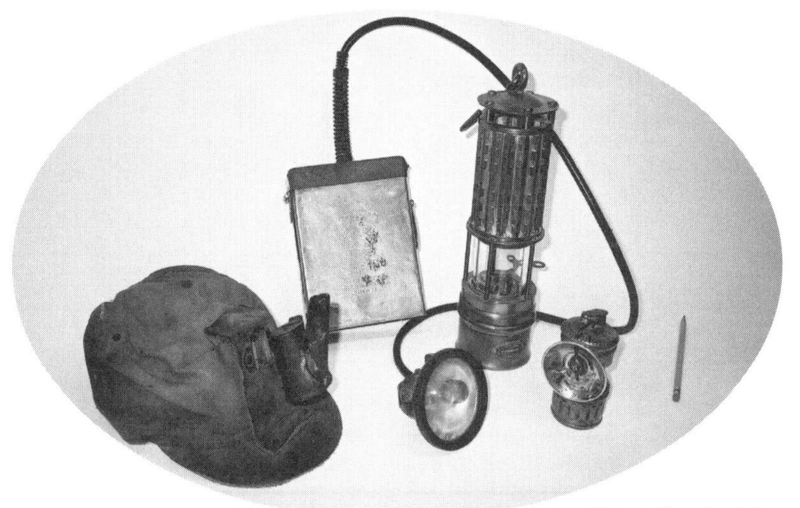

Richard Slingerland Photo

Miners' lamps from left: open flame lamp on a soft miner's cap, battery and cable of an electric lamp, a safety lamp, and a carbide lamp.

Cumberland Museum C160-078

A coal trip emerges from the Number Five South Wellington portal, which is under the main line of the E & N.

Cumberland Museum C160-002

In a Cumberland mine, Japanese miners with lamps glowing work two seams of hard coal, which are separated by dirt or soft coal.

Cumberland Museum C290-002

A colliery locomotive chugs by Union mine camp. More than a century later the stumps are gone, but the homes and gardens remain.

Cumberland Museum C030-007

Off-duty Cumberland miners belly up to the bar at the Vendome Hotel's "Bucket of Blood."

Cumberland Museum C162-007

Outside Number Four Cumberland, drivers, some with whips draped across their chests, pose with their mules.

Cumberland Museum C164-032

In a company publicity photo, two trips of coal specially washed for the occasion, emerge from the Number Four Cumberland portal.

BC Archives 41770

In Cumberland's Chinatown, three thousand Chinese, mostly men, live and work in the mines for low wages, and dream of returning to China.

Cumberland Museum C166-009

A miner works on his knees under a low roof as he begins the horizontal cut necessary to set shots for blasting a narrow seam.

Cumberland Museum C140-304

Japanese miners in a studio pose wear their soft caps and hold their lunch buckets, but their oil lamps are nowhere in sight.

Notman Photographic Archives, McCord Museum of Canadian History, Montreal View 8901

The Hamilton Powder Company wharf at Departure Bay waits for the ships that bring in nitroglycerine and take away black powder and dynamite.

Notman Photographic Archives, McCord Museum of Canadian History, Montreal View 8909

Though immaculately clean, Number One Hall with its dynamite packing machine is just as dangerous a place to work as any part of the Powder Works.

Cumberland Museum C160-040

Two miners operate a longwall machine as it carves a two-metre-deep horizontal cut into a narrow seam that lies between two layers of rock.

BC Archives 18725

Number One Nanaimo, the oldest and largest operating mine in B.C. in its day, commands the harbour and the mine beneath.

Nanaimo District Museum 03-25

In the 1890s, a tugboat guides a scow that brings lunch-bucket-toting miners home from work in Protection mine.

BC Archives 64282

A ransacked strikebreaker's home in Extension is a victim of the riots of August 13, 1913.

BC Archives 93329

By the late afternoon of August 13, 1913, the Extension pithead was on fire; by morning all that is left are metal skeletons.

Cumberland Museum C341-006

Escorted by bayonet-toting Seaforth Highlanders, provincial police and "specials" in plainclothes march into Cumberland.

Nanaimo District Museum O5-5

Close to St. Paul's Anglican Church and the Windsor Hotel, recently damaged by the S.S. *Oscar* explosion, the militia camps in Nanaimo's Dallas Square.

BC Archives 8979

Wearing leather aprons to hide their sporrans, Seaforth Highlanders pose along the E & N track at their Prideaux Street camp in Nanaimo.

Ray Knight Collection

Soldiers of the 5th British Columbia Regiment Canadian Garrison Artillery escort Ladysmith strikers up First Avenue while women and children keep pace.

BC Archives 11701

Officers of the militia converse lightheartedly while their commander, Lt.-Col. A.J. Hall, smiles for the camera on the train station platform.

Cumberland Museum C110-001

In Cumberland, miners carry Ginger Goodwin's coffin down Dunsmuir Avenue while in Vancouver, workers stage the first general strike in B.C. history.

Ray Knight Collection

Coffins of the men who died when the Protection cage dropped are carried up Nanaimo's Bastion Street to the cemetery on wagons.

Nanaimo District Museum

Bound for the annual Miners' Picnic, the tugboat *We Two* nudges the scow *Rainbow* toward Newcastle Island.

Ray Knight Collection

A man trip descends the notorious "killer slope" at the change of shifts at Granby mine.

George High Collection

In Number Five Cumberland, rope rider George High, eighteen, shows how to ride a trip of coal cars.

Nanaimo District Museum 03-13

In the 1920s, Lantzville miners use carbide lamps while all other Island miners have converted to electricity.

Cumberland Museum C169-009

Lime rock dust covers the floor of Number Five Cumberland as black-faced miners with modern lunch buckets take a break.

Nanaimo Museum 03-45

A charobanc or "charobang" makes it possible for miners to live in one coal camp and work in another.

Robert Johnson Photograph

In 1982, concrete pillars identify the Million Dollar Mystery whose capped shaft is marked by the stump of a steel girder sold for scrap.

Chapter V

In the Shadow of Number One

Unlike Extension, South Wellington, and Cumberland, Nanaimo has survived its days as a coal camp to become a modern industrial and distribution centre. But for those who know where to look, there are still vestiges of the days when coal was everything. The commemorative chunk of coal on Front Street is the most obvious; the carefully restored miner's house on the grounds of the museum is another. The Rattenbury-designed courthouse, which witnessed the arrest of striking coal miners on an August night in 1913, still looks out over the harbour. And driving about the city, a person who knows where to look can spot the slack heap through the trees on the edge of Diver Lake and can tell which houses in Northfield, Wellington and South End were miners' houses before they blended into the modern scene with added rooms and stucco siding.

The most obvious reminder of the days of coal, however, is gone. When the tipple of Number One Esplanade mine came crashing down in 1950, the symbol of Nanaimo's and indeed the Island's coal mining industry disappeared from sight. But not from memory. Before it ceased operations in 1938, it was the oldest operating mine in B.C. and the largest. Its silhouette was familiar to thousands of people who had lived and worked under its shadow. The soot from its coal-fueled boilers had blackened everything in its immediate vicinity. Homemakers on washday kept a wary eye on the wind. A shift could bring soot raining down on freshly washed clothes drying on backyard clotheslines.

Number One was a part of everyone's life: not just the coal miners working the seven miles of tunnels honeycombing the rocky foundation of the harbour; not just the Chinese boilermen living in the wood-frame ghetto on Pine Street; not just the women for whom sooty clothes were minor irritants compared to the spectre of widowhood. Number One pithead dominated the harbour view of every resident of Nanaimo be they merchant or miner; the economy of the town rested on the thousands of tons of solid black energy blasted from beneath and beyond the harbour, and the whistle ruled the comings and goings of every citizen of Nanaimo and of people as far away as Extension and Second Lake several kilometres away—anyone who could hear its raucous voice.

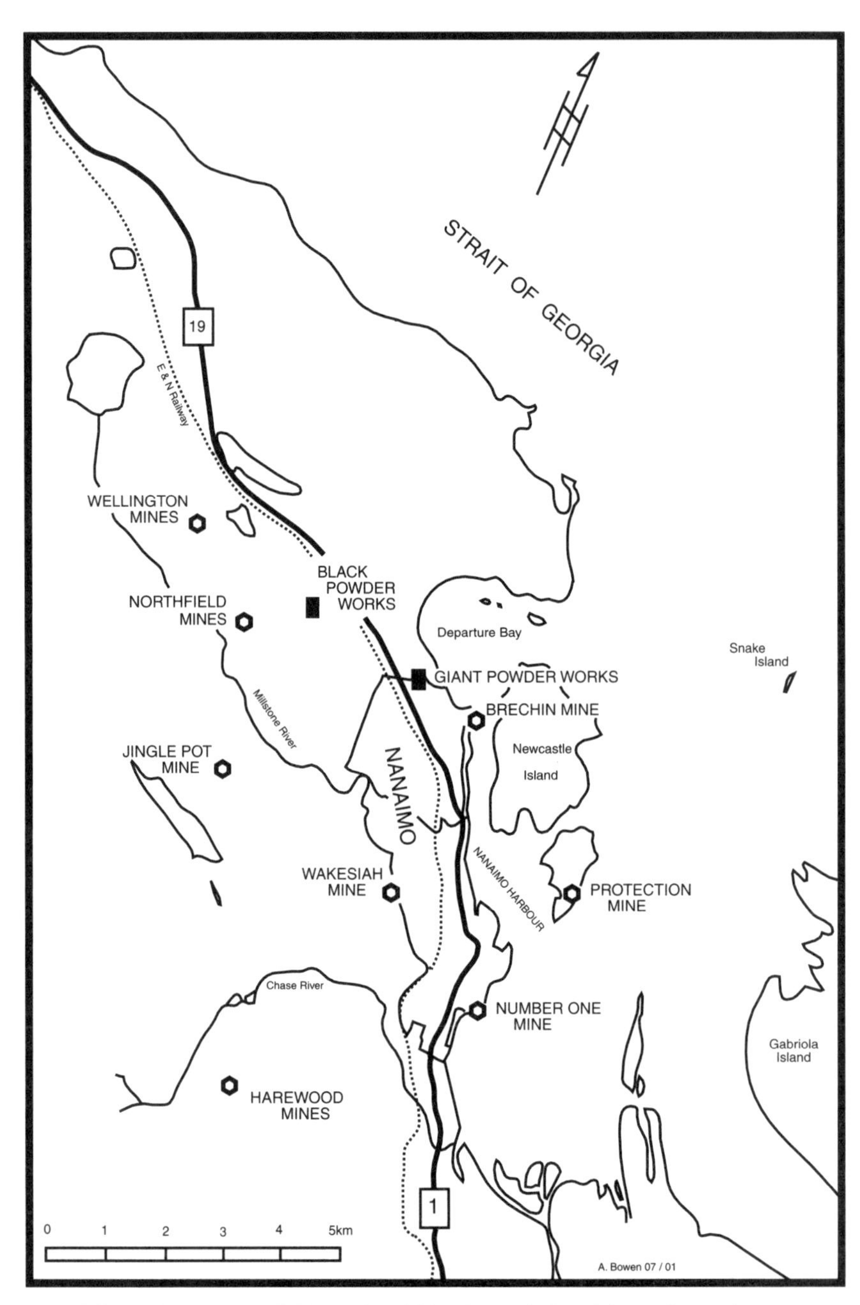

The Mines and Powder Works of the Nanaimo Area

The loud, low-pitched voice of Number One sounded first on November 15, 1883, one month after shaft sinking operations, begun in 1881, had finally tapped the Douglas seam. The owner of the big new mine was the Vancouver Coal Mining and Land Company (Nanaimo Coal Company) whose board of directors lived and made decisions in London, England. The members of the board and the machinery for the mine came from the old world; the bush-hewn timber that shaped its distinctive pithead came from new. On May 3, 1887, the big mine took 148 lives.

> It was five minutes to six in the evening when we heard the thunderous sounds from our house. The ground shook like it was an earthquake, bricks fell down off people's chimneys, and everybody ran screaming, "What's the matter? What's the matter? There's an earthquake or something." Oh they were frightened, running, crying, kids crying. The commotion was terrible.

The explosion touched the lives of everyone in Nanaimo, and Nanaimo still remembers every May 3rd by flying the flag on the HBC Bastion at half mast. And people still tell stories of the explosion that have come down to them from parents and grandparents who were there.

> When my mother was a girl she said she would never forget hearing the continuous whistle from Number One being blown. That meant there had been an accident. She and the rest of the family ran out into Needham Street where they saw a tragic sight. Men, women, and children, some of the women pregnant, streaming down toward the mine. It registered on her mind that so many of the women had starched white aprons on. In those days that meant dressing up for dinner. The women in most cases were tugging little children with tears rolling down their faces.

> Some of the victims were killed outright. They were walled off in a part of the mine. So they wrote farewell messages in the dirt on their shovels. My grandfather and my uncle were both killed in that explosion.

One hundred forty-eight grandfathers, uncles, fathers, and brothers – everyone on the evening shift except for seven men – died either in the explosion or from the afterdamp. The wives, children, and off-duty miners gathering at the pithead to wait for news – it was a scene that has been repeated over and over again in the history of coal mining. When management considered labour expendable, when miners had to choose between being safe and getting out as much coal as possible, when open flame lamps lit poorly ventilated or dusty mines, an explosion was inevitable. Usually the death toll was lower and usually the survivors returned docilely to work in the same conditions as before. But this time the number of dead was huge, and this time there was a second explosion eight months later in a Dunsmuir mine nearby. This time, the surviving miners turned their shock into anger and a demand for action.

They took their demands to Samuel Matthew Robins, Superintendent of the Nanaimo Colliery, who had come to the Island from England in 1884 to supervise the opening and operation of Number One. Although he had a reputation for treating his employees fairly, he at first refused to give in to their demands that Chinese miners be prohibited from working underground. But Robins was no match for the fury the explosion had engendered. He capitulated and decreed that from 1888 onward, in all mines owned by the Nanaimo Coal Company, Chinese and Japanese men would work only above ground.

When Robins retired in 1903 he was much respected for his fair-minded ways. He had shepherded the company through the aftermath of the terrible explosion, through a reorganization in 1889, and through the recession of the 1890s. It was during that troublesome decade that he had allowed the Miners' and Mine Labourers' Protective Association, a local union, to represent the miners, to present their grievances to management and negotiate layoffs in a way that made it easier for the miners to support their families.

Robins took his leave of Nanaimo at a celebration held in his honour on The Green, a picnic area near the harbour. A large crowd came out to say goodbye to the man who had also donated a park to the Cricket Club to be used for horse racing and other sporting events and the man who had made it possible for mining families to acquire two-hectare homesteads.

Robins had set the newly jobless Chinese to work clearing heavily treed company land between the southwest boundary of town and the foothills of Mount Benson. Two hectares equal five acres in imperial measurements, which were in use at the time, and hence the district came to be called Five Acres. On the condition that the land be fenced and cultivated, miners could lease land at minimal cost with an option to buy. The plan enabled miners to raise their own food and support their families even during mine layoffs.

But as popular as the Five Acres scheme proved to be, Sam Robins was most popular for his running feud with the Dunsmuirs. Stories of his encounters with the unpopular coal barons take on an apocryphal tone from constant retelling.

> There used to be a fence on the boundary of the Hudson's Bay Company land from Departure Bay up to where Madill's shop is now. So Robins' company took over that land grant and Dunsmuir's coal lease was right beside it. Well when Robins and Dunsmuir started rowing they got suspicious of each other and there was a guard put on at night to see that they didn't shift the fence over.

> When old Robert Dunsmuir put the E & N Railway through, Sam Robins needed a level crossing from Northfield mine. Dunsmuir told him that there was his railway and they could cross either over the top or underneath but they couldn't have a level crossing. That meant they had to change their grades and backswitch all the way down to the ocean. So a few years later, James Dunsmuir wanted to start the Extension operations and he planned to ship from his Departure Bay shipping facilities. It meant he had to get Robins' permission for a level crossing of the Northfield line. Robins said, "There is our railway and you can either go underneath or over the top but you can't have a level crossing." So that's when Dunsmuir decided to ship at Ladysmith.
>
> The old Southfield mine was being dismantled and Sam Robins approached James Dunsmuir and said he wanted to take out a huge pillar of coal which was under the E & N but that he would leave in props. Dunsmuir wouldn't cooperate so Robins told the men to take the props out. Every time the train went over, the miners would scurry for cover. Finally one day the whole thing caved in. There was a hole about ten feet deep and fifty or sixty yards wide. No one was hurt although the train passengers had to walk on planks over the cavity to get a new train which was brought up from Victoria.

Some people say that Robins allowed his feud with the Dunsmuirs to cloud his good judgment and that his image was shiny more by contrast to the Dunsmuirs' than by actual merit. They say that he was too paternalistic and that the miner's union was a tool of the company and not a real union. But there is no disputing his contributions to Nanaimo, and the city feels his presence still in the row of Lombardy poplars that lines the former site of the company farm on Wakesiah Avenue and in the trees brought from all over the world by collier captains that still grow in his former garden on the Esplanade.

After shepherding the transfer of ownership of all the company holdings to a San Francisco-based group known as the Western Fuel Company in 1903, Robins returned to England. The new manager, Thomas Stockett, assumed control of Number One mine with its shafts on the Esplanade and on Protection Island and its 1,300 employees. When fire destroyed the newly built head gear and pithead on May 29, 1904, Stockett supervised the building of a twenty-seven-metre high pithead, designed by an architect from San Francisco, and equipped it with the latest in mining machinery.

Chinese workers hand-fired a battery of boilers to provide the steam that drove two huge air compressors with fly wheels 3.5 metres in diameter. Compressed air piped through the mine ran rock drills and winches for hauling coal. Steam also drove three large engines: the fan engine pulled air that had entered through shafts on Protection and Newcastle islands. The generating engine produced 250 volts of electricity on heavy cables strung along the roof of the main tunnels to drive the electric locomotives or motors. The third engine, a survivor of the fire, was bolted back together and used for many more years, this time housed in a ground-level building. These engines required so little maintenance that they were stopped for half an hour only once every five years to renew the packing.

Thirty-centimetre-square timbers lined the 182-metre shaft, and the finest equipment in the world lowered and raised men and brought the coal up. Two cages suspended by 3.75-centimetre wire rope from the large drum of the hoisting engine balanced each other. The bottom cage took on loaded mine cars while the top one received empties. At a signal from both locations, the operator engaged the engine. Twelve and a half revolutions of the drum brought the bottom cage up the shaft and the top cage down. The entire procedure, including loading and unloading, took twenty seconds and made it possible to lift nine hundred tonnes of coal in an eight-hour shift.

At shift change, the cage transported men. Sixteen men stepped aboard the steel cage at the bottom, and sixteen at the top. The cagers did a head count and shut the heavily screened metal doors. When the engineer heard one ring from the bottom and two from the top, he engaged the engine. As sixteen men rocketed up the shaft, sixteen men plummeted downward.

Electric lights illuminated the area at shaft bottom where the animals of Number One mine – mules, horses, and Shetland ponies – lived in whitewashed and pristine stables, "good as anybody's house." In the vast network of tunnels that made up Number One and Protection mines there were about three hundred of these haulage animals. Many had names and distinct personalities; the men who drove them remember them well: Shetland ponies like Queenie, Mabel and Jeannie for the low work area where a mule couldn't go; mules like Black Jim – big and black and Roman-nosed – and Buster – Buster by name and Buster by nature – were ornery animals who turned drivers into madmen. Another mule named Fox turned out to be a friend in need.

> Everyone was afraid of Fox, even the "star drivers" which I wasn't. When they told me to take him, I thought if they can't handle him how the heck was I supposed to? But the darn mule saved my life. I was ridin' on an empty car behind Fox when the gol darn car jumped the track. Jammed my knee up against a post. I couldn't reach down and unhook the car. The mule looked back there, switchin' his tail, his long ears wigglin' and I thought he was goin' to go forward and take my leg off. So I said, "Come on, back up Fox. Back up." He'd wiggle his tail. "Back up." Wiggle his tail and back up a bit. Finally he got his back up against the car and he *pushed* it back. He did. And I got free. After that, that mule would do anything for me.

The loaded cars eventually ended up on the scale at shaft bottom. Each car had a tally – a copper disc with a number on it – hanging inside near the top to identify which miner had loaded it. The weighman noted the weight and marked it on the tally sheet beside the miner's number. The number of tons appearing beside the miner's name determined his pay, but the accuracy of the numbers was questionable.

> So maybe we're coming off shift and you'd put your head in the weigh house and look at the tally sheet and they'd all be 1,600, 1,600, 1,600, You knew there was no way they could actually weight them properly. The cars were just going right over.

When it came to detecting rock in a carload of coal, however, the measuring seemed to become more precise, often to the detriment of the miner.

> The cars generally weighed around fourteen hundredweight and if there was a hundred pound of rock you got nothing for the other thirteen hundred. You got nothing. There used to be a board on Protection Island and your number would be on there. You were suspended until you saw the manager. Your job was hanging on that. He'd talk to you like a school teacher spankin' your hand. He'd say things like they was sellin' coal not rock. You would have to make all the excuses in the world otherwise you'd be fired but you still didn't get nothing for that car.

Despite its flaws, the tally system did a good job of keeping track of the miners – where they were working, where they were at shift end – and of the equipment. On a board outside the lamp cabin hung a little brass check for each man who was employed underground. To get a safety light, a miner took his number off the board, handed it to the lamp man, who gave him a lamp with the same number. At the end of the shift the miner returned the lamp in exchange for the brass check which he hung on the board outside. A quick glance would tell if anyone was missing.

> If you didn't come back out, they knew there was somethin' wrong and they'd go searching for you, because you could get lost in the mine too, in these old workin's you know. If a man was gonna get inquisitive, goin' around where 'e's not supposed to go, he could get lost or hurt. When they brought a miners' body out after an explosion, they'd put the check number on his boot.

The lamp man's job carried a measure of responsibility but very little danger. But one of the lamp men working for Number One mine in 1927 would soon have danger associated with his very name. A quiet man of small stature, Edward Wilson had arrived in Nanaimo in April in the company of two women. The trio set up housekeeping in Northfield and Wilson got a job as lamp man at Number One where he was known for his efficiency and for the fact that the lamp cabin was full of books which he read during the long lulls between the beginning and end of shifts.

Wilson did not stay a lamp man for long. Soon he purchased land south of Nanaimo with money given to him by wealthy disciples attracted to his colony which, he predicted, would be the only place to survive the impending end of the world. His Aquarian Foundation would founder within six years amid the publicity of a trial and rumours of slavery and buried gold. By then the former lamp man would be known as Brother Twelve, "Canada's False Prophet."

The miners had known Wilson in another guise too. When he performed as a hypnotist, there was standing room only at the Orpheum Theatre. In the morning after a performance, when the miners collected their lamps and headed off into the maze of tunnels and crosscuts of the mine, they would shake their heads over the things he'd made them do on the stage in front of their friends.

The tunnels and inclines of Number One and Protection mines extended for a total of twelve kilometres. Many of the areas had names rather than numbers: Kileen Incline, Big Incline, Spear's and Lamb's, Cobble Hill, Puywallup. Ha Ha was the incline right off shaft bottom named because an inspector said, "Ha. Ha. You'll never get coal there." The miners had the last laugh.

The tunnels extended towards Protection Island which served, after October 1, 1905, as the entrance for all the men working on the Protection side of the harbour. Number One level, 364 metres below the mud flats, extended towards the Nanaimo River estuary. In Number One north level, two motors each pulled seventy cars of coal at a time. A shaft on Newcastle Island, besides providing another inlet for fresh air, had a ladder for emergency exits.

This network of tunnels, slopes, and inclines tapped a large part of the enormous body of coal in the Nanaimo area estimated in a 1914 geological survey to contain about nine hundred million tonnes in three seams: the Wellington, the Newcastle, and the Douglas, closest to the surface.

Containing highly volatile bituminous coal of fair quality, the seams were irregular and tended towards pocket formation, factors that made it expensive and dangerous to mine. Number One and Protection tapped the Newcastle seam – 2 to 3 metres thick 250 to 300 metres below the surface – and the Douglas seam – 1.5 metres thick between a regular roof and an irregular floor 3 to 30 metres below the surface.

In 1904, the pithead of an entirely new Western Fuel mine appeared on Pimbury Point, Departure Bay, where the channel between Nanaimo and Newcastle Island empties into Departure Bay and where the B. C. Ferries now dock. Number Four Northfield mine tapped the north field of the company's lease and connected with the two seams of the old HBC Fitzwilliam mine on Newcastle Island.

Most of the two hundred miners brought from the north of England to fill the need for experienced men called Number Four Northfield the Brechin mine. Around the miners' cottages that ran in a double line up the hill behind the mine – some newly built, some moved from the old Northfield Number One mine further inland that had closed when the San Francisco market collapsed in 1894 – the homesick men noticed bracken. This fern-like plant, so common on Vancouver Island, was familiar to them. "Bracken" became "Brechin," or so the story goes.

Because Brechin mine needed even more men, there were jobs for Nanaimo miners too. A road that led eventually to Departure Bay cut through the trees hugging the channel that separated Newcastle and Vancouver Islands. Nanaimo miners, some in horsedrawn sulkies but most on foot, went to work past the only signs of civilization: a Japanese fishing settlement on the water side and the Peck Brothers Hotel on the other. The three Peck brothers offered liquid hospitality to Brechin miners as they returned to Nanaimo after work. It was rumoured that Brother Twelve had stopped the Peck brothers from drinking with the power of his eyes.

Brechin mine offered modern equipment, electric lighting, a minimum of gas, hard, bright coal, and easy access to markets: the coal poured directly into ships tied up at the wharf beside the tipple. The mine's two parallel slopes, one for men and animals and one for coal, led down at an angle under the channel towards Newcastle Island.

> My first job was runnin' winch in Brechin mine and I was goin' down the slope and I was never so scared in all my life. 'Cause it was under the salt water and water was comin' in. Couldn't come in any quicker. And I thought I'll never come back out. And I was scared every day I was goin' down there.

Working in a mine beneath the ocean did not bother most of the men employed by Western Fuel in its harbour mines. And those who were bothered could work in the company's Harewood mine – one of the few mines on the Island above sea level.

Situated on the southern side of the Mount Benson foothills, close to the Five Acre homesteads, Harewood mine had a long but sporadic history. In 1863, a company involving Dr. Alfred Benson, for whom the mountain was named, and the Honourable Horace Douglas Lascelles, seventh son of the Earl of Harewood, hired Robert Dunsmuir to open the mine. In 1876, when surveyor T.A. Bulkley owned the mine, he built an elevated tramway that became by far its best-remembered feature. The tramway used buckets to transfer coal from the mine to the waterfront.

The antics of Captain Jemmy Jones, skipper of the first ship to take on Bulkey's coal, are part of Island coal mining lore.

> There was a man who used to ride up to Harewood from downtown by stepping into one of the buckets on the tramway. One day as he was riding up it stopped for twelve o'clock for lunch. A crowd gathered around to see what he'd do and they say he made a speech which lasted one hour until the tramway started again. They said it was a fine speech he made about mine conditions and one thing and another. Some people threw rocks at him and he threw coal back at them. Some said he did it on purpose just to get attention.

The Harewood mine tunnel ran straight into the mountain with slopes angling upward and downward from it. The winch hauled coal up from one side of the tunnel and pulled it down from the other side. The coal was dirty and the mine was wet and dangerous. Its strong roof would not cave in, and that should have been an asset, but instead it proved to be a killer because when it moved, as it inevitably would, it came down in one piece and no amount of timbering could prevent that.

> They leave coal in these old mines and it's usually for a good reason. It's too dangerous to take it out. We thought we could timber a lot and get the coal out but we couldn't. That mine killed my father. The timber just broke and splintered and it hit him. So I just closed it up and walked out of there. I says, "I can find a better place to work than this."

Everyone who tried the Harewood mine learned the same thing. After brief openings in 1876, 1892, and 1902 and a longer period from 1917 to 1923 when First World War subsidies made it pay, it was abandoned. Because its workings were above sea level, it did not flood, but this made it more dangerous. It became a place for children to explore. In 1968 two boys sat down to rest on the floor of the tunnel and weren't seen until another group of children found their skeletons ten years later. The old-timers could have told them all the ways an abandoned mine can be dangerous: explorers can become lost in the maze of levels and slopes; methane gas that bleeds out of the coal can explode; tunnels can collapse; low-lying carbon monoxide gas can induce sleep – sleep from which they will never wake.

The foothills of Mount Benson held another mine whose importance is greater than its brief period of operation would suggest. The Jingle Pot mine was the only mine to settle with the United Mine Workers of America during the Big Strike. But before the strike, the mine, which had belonged to the Dunsmuirs since the 1880s, had been leased to a company whose president was the subject of much speculation in Nanaimo. Some of the speculation involved a Scot named Henry Auchenvole, who worked in the Canadian Collieries office at Union Bay, who may or may not have had anything to do with the Jingle Pot mine, and whose unusual name was often mistaken for one of German origin.

> Canadian Collieries and Western Fuel, it was to their advantage to keep the little fellow out if they could. Buy him out, close him out. But Von Alvensleben, he came and started that mine in 1907. He was a German, a man of money from Vancouver by way of Seattle.

> Von Alvensleben. Yes, he had this Jingle Pot mine and there was another German fellow. Auchinvole. I think he took over his interests. But the last we heard of Alvensleben was he ended up in the States. Broke. But people said it was the German Kaiser's interests he was representing.

> There was a German fellow here and he had a finger in a lot of things around B.C. Loggin' camps. I know he had a pile of timber. And when the war come on, of course he was locked up. Locked up in a stone jail they had down there. That was the last use they made of it for a while. And there was old Auchenvole and he was the man who looked after the office and that.

The mysterious Alvo Von Alvensleben appears in the records in 1908, as the vice-president of the Vancouver-Nanaimo Coal Mining Company in the second year it operated the Jingle Pot mine. A year later he became president, a post he held until 1913. In that period the mine's workforce grew from 24 to 343 employees.

But the president was not what he seemed. The handsome, aristocratic German national, who lived in a mansion in Vancouver and mixed with all the right people, may have been a spy. He may have been involved in a plot to destroy bridges over the Pitt River east of Vancouver. He may have lured Premier Sir Richard McBride's secretary into becoming a spy too. Though it is unlikely that he was confined to Nanaimo's old stone jail, the elderly structure having been pulled down in 1893, Von Alvensleben did disappear after he fled to the United States.

The mine was never the same afterwards. The decline in coal markets at the beginning of the war was especially hard on the old Jingle Pot. Management cut staff by a third. Fires in 1915 and 1917 made walling off whole sections of the mine necessary. A new Jingle Pot mine, also called the Westwood mine and owned by the British Columbia Mining Company, was 450 metres to the west and tapped the same body of coal. After 1920, the new company leased the old mine from Canadian Collieries, pumped the water out, and began to extract coal.

The men who worked the Jingle Pot mine remember the twin slopes too steep for a man to walk up under his own steam, and how the miners were supposed to hold on to sticks attached to a rope and walk up while a car engine pulled on the rope. Later a "man trip" – empty coal cars for the men to ride in – replaced the car engine and rope with sticks. The miners remember the day the cable on the last coal trip of the day broke.

Ten cars of coal, four more than there should have been, hurtled back down the slope and killed the man at the bottom. They remember that the company adopted a double rope system to prevent such runaways. A rope rider remembers how the long 367-metre slope lined up perfectly with the late afternoon sun and how he had to wear goggles so he could see and how funny it must have been to see a man come out of a coal mine wearing sun goggles. But no one knows for sure how the mine got its name.

Some people say there used to be a tripod at the bottom of that long slope with a pot dangling from it. When a lone worker wanted to come to the surface, he would jingle the pot and the rattle of the stones would signal the winch man to pull him up. The Ministry of Mines Report for 1889 mentions a mine that was "facetiously called the 'Jingle Pot'". Such an early reference seems to give credit to the story that the mine's name derives from the custom in certain areas of Great Britain for miners to hide the coins left over from their pay packets in the family's special-occasion teapot. When the family needed emergency money, the miner would ask his wife, "Is there any jingle in the pot, love?" A jingle pot came to mean good fortune and thus may have seemed to some miner the perfect name for mine with a rich coal seam.

The last days of the Jingle Pot mine were uncertain ones. Canadian Collieries would terminate the lease if there was nobody on the job for forty-eight hours.

> The timekeeper lived in an old brick house out there, which was the office at the time. So what they done this Christmas, they got some friends of his to go out there with two bottles of whisky and got him drunk and took him away and didn't bring him back until New Year's. So the lease terminated, the mine was finished. That was 1928. Everything is in that mine. Only thing that come up was the mining machine. So my father and I pumped it out in 1936 and opened it up and we worked it. We took the last coal out of it.

Immediately adjoining the Jingle Pot lease, on the Western Fuel Company farm, ninety-seven-metre twin shafts gave access to the Wakesiah mine. For thirteen years, from 1917 to 1930, the company worked the irregular seam, its measures tilted and thrown up in an exaggerated version of the Vancouver Island pattern. One explosion in 1922 killed Alfred Odgers and Gilbert McBroom. Like many other mines, the Wakesiah died for lack of markets, and its shafts and tunnels filled with water.

In order to sink the shafts and work the tunnels of any mine, miners needed blasting or gunpowder, a cheap compound suitable to use in coal mines because of its slow action and the relatively low level of danger involved in its use. Blasting is not a safe procedure, but blasting with powder in compressed cartridges or cylinders by properly trained and conscientious miners in well-ventilated mines is as close to safe as a mine can get.

Nanaimo needed a powder works in the nineteenth century because the cost of shipping blasting powder from England was high, and the powder arrived sodden with salt water. Accordingly, in 1887, the Nanaimo Coal Company had encouraged the Hamilton Powder Company to establish a black powder works on three hectares of donated land outside the city limits, in accordance with the law, on the present site of Beban Park. It happened to be near Northfield Number One mine, which had just opened.

When the powder company decided to manufacture dynamite as well, however, it added a much more dangerous dimension because the process required nitroglycerine. A chemical plant in Victoria agreed to supply the volatile yellow liquid. Huge glass demijohns fitted with wooden collars came by boat to Departure Bay. It took two men, moving very carefully, to lift each of these monsters into a horse-drawn wagon for the trip to Northfield, three or four kilometres away. Extra heavy springs and ten centimetres of wood shavings on the wagon floor gave the six to eight glass containers a small amount of cushioning. A lone man drove, his back to the ominous cargo.

Black Powder Road was a steep trail that came up from the dock and continued rutted and muddy until it reached Northfield Road, which wasn't in much better condition. The wagon driver knew how dangerous his job was. He had done it many times and he had always insisted on doing it alone. His luck held until one day in 1896 just when he was crossing the E & N tracks.

> One morning it blew. They don't know exactly what happened, maybe two of those big demijohns started to rub together. But the horse and the wagon and everything went and him right in the middle of it. He was blown to bits; the horse was blown to bits. They never did find him or the wagon. Made a hole in the road there you could sink a house into.

Not long after the explosion, the company built its own giant powder works on land sloping down to the ocean at Departure Bay. They would manufacture their own nitroglycerine and make the dynamite there and so eliminate the hazardous journey to the Northfield plant, which would continue to manufacture blasting powder.

Present-day residents of Cilaire, a well-manicured enclave of expensive homes, each with an ocean view, might find it difficult to picture the same property in the early years of the twentieth century. Two different processes – one to manufacture dynamite, the other to produce gelignite – occurred in a series of buildings that began at the top of the hill with the nitrator and progressed down towards the ocean. The separator, recovery grist mill, wash house and storage, mixing, packing, and rolling houses all were connected with wooden sidewalks. In each of these buildings, Chinese and white men and young girls, favoured for their delicate touch, laboured at a harrowing occupation. At any moment, without warning, they could be blown to bits. Each person's life depended upon how carefully each other worker did his or her job.

> You had to be real careful. No metal allowed. The shovels were wooden and things were made of pressed paper. You couldn't go into them buildings at Departure Bay with your shoes that you were walkin' around in on account of the nails in the bottom could cause a spark. They always had big shoes for you to slip on.
>
> One Sunday some carpenters had been workin' in the mix house doin' some repair work and I went in there and I seen this big parlour match on the floor. They had an awful big head on them. If it had ever went off...
>
> We weren't afraid. The only thing, when you first started there you would get a terrific headache. It would pret' near drive you crazy. After three days you started to lose it and you never had a headache again.
>
> Oh, I guess we figured it was nicer than working down in the coal mine.

On January 4, 1903, at 7:00 a.m., twelve men employed in the mixing and rolling houses making gelignite were "launched into eternity," as the local paper described it in a large headline. Although one of the safest explosives to use, gelignite was very difficult and dangerous to manufacture. No one ever determined the cause of the explosion, but its effects were immediate and awe inspiring.

> They must have had quite a bit of stuff lying around because it was a terrific explosion. People said they heard it in Parksville and Ladysmith. Windows all over town were broken. I was just a kid and I had just got out of bed. I was in the kitchen. There was a lamp on the table and the explosion shook the lamp chimney. My mother was in the other room and she thought I was playing with it. In them days they used to warn kids to keep away from the lamps. Just then, we looked out and there was all the women runnin' down the street to the mine. They thought sure there was another mine disaster. Some of the women were pullin' their hair out.
>
> There was ten Chinamen and two white men got blew up there and the biggest piece we found was one man's arm. Everybody went around with buckets and pails and sticks to pick up the flesh and bits of hair. There was a six-foot-deep excavation. A railroad came in to the powder works there and it blew up one of the rails and wrapped it around a tree. It bent just like a snake around that tree.

The explosion made the residents of Nanaimo realize that the powder works put the whole town in danger. But as the years passed and nothing more happened, people felt more secure until the separator building blew up on May 19, 1910, taking five men with it. Eyewitnesses told of a sound like a thunderclap and a "mass of earth and debris which fell like hail all around the country, up to half a mile away." Dense smoke settled over the bay, and all that was left was a "yawning chasm with a seething, foaming, steam-like vapour, hissing and ominously roaring." There were only two caskets at the funeral: one contained a body, the other contained all that remained of the other four men.

By 1911, the powder works had a new owner, Canadian Explosives Ltd. An explosion on October 25th didn't kill anyone. Then, two months later, just three days before Christmas in the middle of the morning, the gelignite room exploded again.

> They said there must have been over 800 tons of this gelignite went up. And there was no barricades around the buildings then. Three got killed that day, old Joe Depries, and Willy Day, and a fella by the name of Wilcox.

> Those explosions always seemed to take place when the atmosphere was heavy and muggy. I had just taken all this stuff from the grist mill to the mix house and as I walked away I stopped to talk to the man who was mixin' the powder and he said, "It's a bad day, a heavy bad day." So anyway he was on his way to the outdoor privy and after the explosion they found him in there just sittin' with one brace up and there was somethin' hit him in the throat and cut his artery, some glass cut his throat.

> I was blendin' for the gelignite. So I had to weigh out all the gun cotton and the glycerine and mix it up with my hands and then cart it down to the mix house, eh? It was board sidewalks and plank runways and that cart wiggled and shook. Then I went back to the gun cotton house to screen some cotton. I guess I was there about five minutes 'n' I heard a report just like a rifle. Bang.

> And that's all I knew until I come to in the hospital. The whole gun cotton house had come down on me, the whole building.You see the concussion takes the air out and everything falls in and collapses. Well I was underneath all that. I didn't know how I got out of there and I asked Jack. He says, "You didn't. We hadda dig you out," He said there was a splinter of wood right through my jaw that had pinned me down. A splinter of wood.

It was still possible only to speculate on the causes of the powder works explosions. Too much of the forensic evidence was destroyed each time. Yawning chasms and fragmented bodies give only mute testimony. The company applied new restrictions, it built barricades around individual buildings, but the explosions continued. The citizens of Nanaimo were frightened and demanded action, but it took another spectacular explosion seven kilometres away from the powder works to precipitate action.

On January 15, 1913, in the early afternoon, heavily falling snow had reduced visibility in Nanaimo to nothing. The atmosphere was leaden.

Mayor John Shaw stood in front of a large plate glass window in the Windsor Hotel, and though he looked toward the harbour, a curtain of falling snow blocked his view. Suddenly there was a bright flash then a terrible crash as the window shattered, covering the mayor with shards of glass. Outside, people poured out of buildings to escape flying glass only to encounter horses dragging their broken harnesses running loose on the streets.

> We lived on Kennedy Street and you could hear it all over. You felt the air going back towards the mountains and then all of a sudden there was another rush of air coming in and this is what took all the windows out.

> We had a house up on Victoria Road with a fireplace and Mum said, "I wish you'd take the ashes out." Just when I went to do it, the explosion came and all the ashes came out at me.

> When that boat exploded I was a child and it blew out every window in the school. The Harrar girls were school teachers, three of them, and one of them, Pauline, was my teacher and she said, "Children stand up and walk out of here like soldiers. Now march." And we went out of there and it was so wonderful because she was so calm and we were so terrified.

> We was all workin' at the powder works and they had changed to a kind of French window so any concussion and they'd fly open. And that time they worked fine, every window blew open. Well we all ran out and we didn't know where the explosion was. There was lots of scrap coming down. So we met other fellows running around. "Your place all right?" "Yes, is yours?" "Yes, oh yes." A bunch of carpenters there were sure it was the nitrator. "Well," I said, "a big chunk dropped down where I was working but the nitrator's still going." The superintendent come along and he said he'd got a phone call that a boat blew up on Protection Island.

> Now this boat, they called the S.S. *Oscar*, it went to Telegraph Bay near Victoria and loaded 1,910 cases of giant powder into the hold. And it come up here to Departure Bay on January 14th to the powder works and it took on 1,800 kegs of black powder and they loaded that on the deck and covered it over with tarpaulin. Then they went down Newcastle Channel and into Nanaimo harbour to the coal bunkers and took on bunker coal.

Loading was completed and the S.S. *Oscar* cleared for Howe Sound. As it rounded the lighthouse on Gallows Point, the full force of a winter gale struck. Reluctantly, the captain turned back. The S.S. *Oscar* and her dangerous cargo spent the night moored to a dock.

The captain and crew woke to the heavy snow that drastically reduced visibility. Then someone discovered a fire in the bunker coal, a fire so well established that the pumps were useless. The captain made a courageous decision. He instructed the crew to cast off and head at full speed out of the harbour. He planned to order the crew into a lifeboat and then set a course once the vessel cleared the Gallows Point lighthouse on the tip of Protection Island.. Then he would abandon ship himself.

> Well in their excitement the crew lost the lifeboat; it got away on them. So the captain headed for Protection Island and there was a lady over there with two children and she seen this boat coming, on fire, come right up on the beach and they ran out to watch the excitement. And the crew was running off and they grabbed the woman and her kids and they just got under cover when it exploded.

The explosion scattered debris over a huge area, destroyed the boiler smokestacks, pithead, and wharves of Protection mine, shattered windows in town and caused many minor injuries.

Nanaimo's initial reaction was to make jokes. The *Free Press* reported that "sticking plaster is the fashion today and the saloons are doing a roaring business supplying stimulants to victims of shattered nerves." The students of the Old Central School got a two-day holiday while workmen repaired windows. The students of Quennel school were indignant because they did not get a holiday, the explosion having "so impolitely passed them over."

When nerves calmed down and victims of flying glass no longer filled doctors' waiting rooms, the feeling of outrage began to build again. One week after the S.S. *Oscar* blew to smithereens on a Protection Island beach, a mass meeting of citizens demanded an inquiry into the manufacture, storage, and transportation of powder in Nanaimo and vicinity. They likened the situation to living with a sword over their heads.

An explosion at the powder works on July 20th of that same year, the unresolved strike in the mines and the rumblings of war in Europe punctuated the demand. On August 8, 1913, Canadian Explosives Ltd. announced a move to James Island off Victoria. It would take two years to build the new plant. Three months later fire threatened the powder magazine at the Northfield Black Powder Works. Four hundred people, convinced that the decision to move to James Island had come much too late, spent a night in the woods.

The sulfur needed to make blasting powder came from Japan in woven sacks, which were dumped on the wharf at Departure Bay, loaded into wagons and transported to Northfield to be stored until needed. The powder magazine was right next door.

> Somehow or other this sulfur house got on fire and the alarm was given and they got a train out ready to evacuate people 'cause they thought for sure this powder magazine was going to go up. And my cousin, he got a bunch of Chinamen there to work for him and they dug a trench. The sulfur will run like water when it's burning. And they drained it away from the magazine and saved it.

That was the last big scare. The powder works closed, the buildings came down, the town breathed more easily, and the complaining started.

> As soon as the plant moved, they started on about the payroll. The payroll wasn't coming here any more. Maybe eighty-five to one hundred jobs was lost. People have short memories.

Family members of former employees still wore the white powder works coveralls. They had no pockets, but the price was right. But soon evidence of the powder works began to disappear. The coal company replaced the Protection Island mine pithead equipment that the S.S. *Oscar* destroyed. The mine had always sent its coal to Number One mine to be hoisted up the shaft at the foot of Farquhar Street anyway. Apart from that, the two mines were separate and differed in a number of ways: Number One held steam coal in wide seams, some up to six metres thick; Protection's narrow seams – sixty centimetres or less – held fine, hard, bituminous coal. Number One used pillar and stall and Protection used longwall in tunnels extending under Newcastle and Protection islands and beyond towards Snake Island in the Strait of Georgia. So cramped were the miners in some areas of Protection mine that they got stuck between the machinery and the roof when working over the pan. The quality of the coal was the only consolation.

> We worked in about a three-foot space. You had to lie on your side. Work shoulder to shoulder. You're always dripping in water. You go down and the first thing you know you're dripping wet. You stay that way and work that way, never getting dried out. It would be dripping from the roof and you'd be lying in the water.

> I didn't like working the low seam in Protection. You lay on your belly and by the end of the day the diggers had so much rock around you, you could hardly get out. When I finally went to Reserve mine it was like heaven because I could stand up.

> The first shift I went on in Protection was wet. When you're fresh from Italy and you don't know how to talk, they put you in a place where the water is coming down and I got no choice. So I said to myself, *The best thing for you to do is take off your shirt and just work in your pants. At least your shirt will be dry when you go home.* So I was there for half a shift until I finished that job and then I put on my dry shirt and then the water come down like buckets.

By virtue of a quality fan and four fresh air outlets, the ventilation system for Number One served both mines well. There was virtually no gas on the Protection side. As the mine workings extended farther and farther out under the Strait of Georgia, the danger increased because there was no return airway and therefore only one way out – back in the direction the men came in. If this was blocked with a cave-in, the fate of the miners behind it would depend on how fast rescuers could get through the rubble.

The nature of the seam changed drastically as the workings approached Gabriola Island. Here the coal was very thick; some said the seam stood as much as twenty-one metres high. Such a seam would have been valuable if it could have been mined effectively.

> They couldn't work it because it used to take fire every time. They had to shield it off to keep the air away 'cause as soon as the oxygen hit it, it would take fire. So every now and then they would open it and work it for a while but they could never stay there very long. It took fire because of the pressure or something like that. Well some places when you used nails to put timbers up, the nails would feel hot. The diggers just wore a pair of football shorts. And they couldn't stay in there long enough to load a car; they had to come out again.

Each time the seam ignited, management closed the area off and allowed it to flood. Even then it was a cause for wonder. Sometimes the water flooding these subterranean tunnels would recede as if it was independent of the forces that affected the ocean above it.

It was hard to ignore the fact that the mines lay beneath up to twenty fathoms of salt water. Some men never could come to terms with it and sought work in other mines instead. They could work under two hundred metres of rock but not thirty-six metres of ocean. But most men who worked Number One and Protection viewed the ocean's presence as a curiosity. They even used the sound-conducting properties of water to act as a time clock.

> The *Patricia* was the passenger boat that went to Vancouver every day and it went out at quarter past two and the boys in the mines could hear it go out overhead. Time to go home now. Get your coat 'n' bucket, 'n'get out. They could hear the drum o' the engines.

Ninety metres to the left of the Gallows Point lighthouse was the wharf and headgear of the Protection Island mine. A cage lowered 16 men at a time 180 metres to their work. This cage was not the sleek efficient machine that its counterpart on the other side of the harbour was. In fact it was a rickety thing that squeaked and rattled at the joints as it descended. By all appearances though, the suspension cable was strong and reliable, constructed as it was from many strands of wire around a hemp core. Every day the cable, cage, safety devices, and wheel on the headgear were inspected. The company's rope expert periodically checked for broken strands by allowing the whole length of cable to run through his hands as the cage dropped slowly. Designed to bear up to eighty-two tonnes, the cable's usual load was only three tonnes. Yet on September 19, 1918, the cable broke midway through a descent with sixteen men aboard.

Six cages each with sixteen men on them had already descended that morning. Ninety-six men were already receiving their assignments at shaft bottom. The entire night shift had already come up to the surface. The next sixteen men began to file on to the cage. Among them was a man still absent-mindedly carrying his pipe. Someone reminded him to leave the pipe behind, but while he walked over to where the smoking things were stored someone else took his place on the cage. A mine inspector gallantly stepped back to give his place to a miner anxious to get to work. With one place left on the cage, a man rudely pushed ahead. The cager stopped another man, shut the door, and sent a signal to shaft bottom.

On hearing the answering signal, the cager started the descent and watched the cage disappear down the shaft. Suddenly an end of cable shot into the air and just as quickly disappeared. The heavy cage had plummeted, hit the bottom, and driven through the thirty-centimetre-square timbers that formed the landing platform. It came to rest nine metres below wedged at an awkward angle, its human cargo crushed in a shapeless mass. The watch on Robert McArthur's lifeless arm read 7:10.

To the men in the mine it sounded as if the roof had caved, so loud was the roar of the cage falling. The manager of the mine, John Hunt, heard the news by phone on the Nanaimo side and made his way through the main motor level to supervise the long and arduous task of extracting the bodies. It was not until late evening that the victims were identified. Fourteen were married, forty children became orphans in an instant, fourteen widows faced the prospect of living on a monthly compensation of $25. Speculation and rumours began immediately.

> You see, the war was on at the time and there was a lot of talk that somebody had filed through the rope and tried to blame it on the Germans and all this sort of stuff. I just think it was neglect on the company's part. The rope had never been examined properly. That's my own opinion.

> The rope broke. I guess it couldn't have been looked after very good. And the salt water running down the rope and drippin' in the mine. Take a pipe in your house if you live by salt water and it's a black pipe. Two years you gotta throw it out. It eats away. And I think that was the same with the rope. The rope broke and them men lost their lives.

At the coroner's enquiry the testimony was both speculative and technical. Witnesses pointed out that despite the regular inspection of the rope, the broken part was always inside the engine house wound on the drum when the cage was at rest. The company rope expert repeated how he inspected every inch of the rope for broken strands. A laboratory test of the rope showed that each strand by itself could hold seventy-three tonnes. The conclusion of the inquiry – the cable broke due to oxidization of the wires exposed to more than normally corrosive water and humid atmosphere – did not satisfy everyone. Even today, people cannot agree on who to blame. When asked by his small son to explain how it happened one man said, "God allows it."

The deaths cast a pall on the town. Just when the First World War was showing signs of at last being over, just when there might be an end to the news of young men dying in the trenches of France, sixteen of Nanaimo's citizens had died in their own home town.

> I saw the sixteen miners from Protection cage come back in sacks. I saw them lyin' on the floors there at Jenkin's old parlour. Two of them were my uncles.

All the stores closed for the funeral. People lining the route of the procession watched as the Silver Cornet Band, with drums muffled, marched slowly, by followed by the mayor and aldermen, then lodge members, then a double line of five hundred fellow workmen just ahead of the hearses, some of which were open trucks carrying three coffins each. Up Bastion Street to Wallace, down Wallace to Comox, along Comox to the cemetery donated by the old coal company, the mine's presence there even after death.

Even when men die, coal mines do not stop work for very long. As soon as the debris is gone, the work resumes. The Protection miners got to work by coming through from Number One, and the blasting, heaving and hauling of coal went on as before. The replacement cage carried only twelve men at one time. Improved safety catches prevented any more cages from dropping, a belated government commitment to worker safety that even proved beneficial to production.

The first concrete evidence of this increased government concern had been the building in 1913 of the first rescue station in the province. At least two company rescue stations already existed: at Number Six in Cumberland and at Number One in Nanaimo, but this rescue station on Farquhar Street near the Number One pithead was the first government-built one. But the men who were most likely to need rescuing, the miners themselves, had taken action to prevent loss of life as early as 1888. Although the CMR Act had long required mines to employ gas committees, it was the miners themselves, tired of waiting for management to act, who paid for them out of their own pockets. The committees inspected for gas in all places in each mine once a month. Each mine took several days and the results were posted where everyone could see them. If deemed unsafe, an area was to be barricaded until it was gas free.

> We paid two fellas two bits every payday from each man's pay and they used to go down these old workings and see if they were clear of gas and every-

thing because they were caving all the time and a bunch of rock comes down and causes a spark and it's that spark that causes the explosion. So the gas committee has to see if these old workings are clear of gas and they'd go and put their report up and the fire bosses that was on weekends were supposed to go through there and double check and they were supposed to mark the rock with their initials and the date to show that they had inspected there.

A government commission enquiring into causes of explosions in 1902 had laid down the criteria for judging an area unsafe: the atmosphere saturated with inflammable gas; roadways, faces and passages dry and dusty; sudden outbursts of gas. In these conditions flame from blasting and defective safety lamps or naked lights could cause explosions. The commission acknowledged that because B.C. was short of experienced miners, accidents were more likely to happen. But despite all this, the commission saw no reason why naked lights could not continue to be used provided the mine met certain conditions: proper ventilation; dry, dusty surfaces wetted down, lamps properly trimmed and cleaned – all these were the responsibility of the company that owned the mine.

Having assigned responsibility to the companies, the commission allowed the continued use of naked flame lamps. It was not until 1918, after sixteen more years of dreadful explosions and loss of life, that naked lights were eliminated from all Vancouver Island coal mines with the exception of Nanoose Collieries mine at Lantzville. It was up to mine inspectors to enforce the government regulations.

Every mine in the province was inspected on several occasions each year and the findings published in a yearly report available to the public. The inspectors were men experienced in mining and on the lookout for infractions of the CMR Act. In theory, the inspection system was a good one. In practice, it had its weaknesses.

The lamp man would say, "The inspector's coming today." Well you know what that meant. It was like a young bride having her first visit from her mother-in-law. She'd be cleaning and cleaning and cleaning. Well, everybody knows the inspector's coming. Every place is going to be timbered up and the company would speed the fan up a bit.

In a mine nothing can stop the coal. The coal has got to go. Sometimes a trip will be coming really fast and maybe a bit of dirt got on the track and it would knock the whole trip off and maybe knock down nine or ten sets of timber too. Well, the boss'd come and say, "Clear the track and get the cars going." Under that bad roof. And it would stay like that until the weekend and then on Sunday they'd get the timbering men to come in and fix it. But in the meantime there were men walking through there every shift and the rope rider running up and down. They talk about safety.

The regulations were good but they weren't enforced. I guess the poor inspector was up against it. If he started reporting on things they'd probably move him way up to Alaska or get rid of him somehow. But the last inspector by golly, the boss was scared of him. He'd stop the lamp man before he could phone down and warn anyone. He'd say, "Don't you touch that phone." And he didn't. He was scared to touch it. So then that inspector he'd walk down and he'd see just how things were. When he retired, he gave me the highest compliment it's possible to pay a mine rescue man. He said to me, "I hope if I ever was entombed in a mine, that you were coming to get me." I thought that was pretty good.

The inspector was speaking to a draegerman, a member of a mine rescue team. Although the law requiring mines to train men in the art of mine rescue had existed since the early twentieth century, it was a long time being implemented. Number Five mine in South Wellington did not have a team until 1934. Building the government rescue station in 1913, however, was a positive step. It was the place where teams from all area mines received training.

And we took the training course and we were well paid. We got $2.50 for every two hours we put in. And then on competition days they gave us a shirt and a pair of overalls. I always said that they were scared some of us would come with the ass out of our pants. And when it came time for competition days we'd go down two or three times a week and train for hours. And we had a wonderful team.

Competitions kept the rescue teams in a peak state of readiness. It is ironic, however, that by the time each mine had a rescue team, the danger of explosions was markedly lower due to improved ventilation, dusting and wetting procedures to control coal dust, and the use of electric head lamps. But the rescue teams were still needed from time to time.

If there was a mine explosion you'd get carbon dioxide and you put the machine on to breathe – they had an oxygen bottle and it was good for half an hour travelling each way. And six men on a team would go down to where the explosion was or the timbers were blown out or the doors had short-circuited the air and they had to do all that repair, to circulate the air and get it going again because they didn't know but there might be men still alive.

When the mine whistle blew the doctor's signal, the whole town knew that someone had been hurt in the mine. A miner was buried under a fall of rock, a rope rider had caught his hand between a car and a timber, a rock man had impaled his foot with a pick. There were no hard hats or safety shoes worn until after the 1950s. The fire boss administered first aid if he knew what he was doing and had the equipment to do it with. Miners would lift the injured man onto the top of a car loaded with coal heading to the surface where the doctor waited.

My partner was my brother-in-law and a great big rock the length of a car come down between us. After that a little piece come down and split his head

> right open. You don't want to get a guy panicky so he asked me how it looked when I took his hat off and I said, "Oh, it's not too bad but we'd better get you out." So I took him to the fire boss station and on the way he put his finger into the cut and then he said he felt really bad.

> My father got hit by a runaway car. Three compound fractures in one leg and two compound fractures in the other. An automatic spring switch didn't open. A drunken doctor saved my father with some brandy. He survived but he didn't get another job. His legs were crippled.

The crippled man received $40 per month compensation to keep himself, his wife, and his seven children. As little as this seems it was a great deal more than he could have expected prior to the establishment of the Workman's Compensation Board (WCB) in 1917. An imperfect organization to say the least, the Board did improve its approach over the years to the point where most workers could expect proper medical care and money to support themselves while they convalesced. Long before the WCB, the miners had been looking after themselves.

Following the pattern set in England since the eighteenth century, coal miners were active in various benefit funds or friendly societies. These ranged in type from the large lodges such as the Ancient Order of Foresters and the Knights of Pythias to the Sick and Accident funds set up by employees of individual mines. For a monthly payment of approximately $3, a member was entitled to a stipend of $1 per day for three to six months when he was sick or injured. There were many examples of these funds and most people belonged to more than one.

Benefit funds were the precursors of most modern social welfare legislation. Not only did they perform the same functions as the WCB, life insurance companies, and pension plans, but they were the forerunner of medicare as well. The employees of each mine contracted with one doctor to provide all medical care and some standard medicines for $1 per month per miner. The $1 fee covered all dependents and all care except maternity. The contracts were much sought after by doctors, a situation which created some unseemly behavior as the doctors vied with each other to be the one chosen by the miners. A successful candidate then had to endure a discussion of the way he practised medicine when his contract came up for renewal each year. Heaven help the doctor who had refused to make a house call.

This was self-help. And the movement went even further. When the hospital needed an X-ray unit, employees of Western Fuel formed the X-ray Contributors Fund in 1926. Each man paid $1, the fund purchased the equipment, members received free x-rays and non-members could buy access to the service. The employee-initiated Hospital Treatment Fund provided the Nanaimo hospital with its largest single source of revenue. In Cumberland, the miners built their town's hospital.

One of the primary reasons for friendly societies had been the universal fear of ending life in a pauper's grave. Consequently, the societies set aside part of the fund for burial expenses. Members received a funeral according to a special lodge ritual, and their widows received money to tide them over the first few weeks. After that the widow was dependent on her family, her husband's fellow workers and her own ingenuity. It is no wonder that the sound of the mine whistle blowing to announce an accident filled the hearts of the women with such dread.

The spectre of sudden death sat over coal mining towns like an ever-present cloud, but everyone pretended to ignore it. A wife never told her husband of her fears and he never told her of his. They prayed it would never happen and prepared themselves for the day when it would.

> It was very important to a miner's wife to send her husband off in the morning properly fed. I never heard of a man getting himself up and off to work. Even if they were sick, the women got up. My father would get up and light the fire and then go back to bed. Then my mother got up when it was warm and did the rest. Give them a good breakfast, pack a good lunch for their buckets and get them off in good shape. It was important to her because she never knew if they were coming back alive or in a sack.

> You always tried to have dinner waiting for them when they came in. I have a friend who lost three men in her family, Her father, her husband, and her brother were all living with her and they were all on different shifts. So they said, "Let's go on the same shift so that mother won't have to do so much cooking." They managed to get on the same shift and the very first day they went down, there was an explosion and they were all killed.

> When he leaves the house you wonder if he's coming back again. Every day it was the same, every day of our lives.

> My father and another man didn't get out of Wakesiah mine after an explosion. My mother was pretty well shook up and very ill after that. We were getting some compensation but it wasn't enough and my sisters and my brothers and I were too young to work. But there were good-willed miners and they helped us. When my mother died, they sent us all to an orphanage, but my aunt and uncle came and got us and they raised us.

Coal mining families needed each other to survive. Many families became related as their children grew up and married within the community. The network grew until each family had several men in the mines. The Storeys were such a family, and when their daughter Ethel married Daisy Waugh, two more more mining families were united.

Everyone who worked in Number One or Protection in the 1930s knew James "Daisy" Waugh. They knew him for his prowess on the soccer field and for his sunny nickname derived from the days when he brought his mother the fresh white and yellow flowers from the daisy field back of Robins Park. He loved parties, and there was no better night for parties in Nanaimo than New Year's Eve. On that night there were dances everywhere, and there was first footing, the good-luck custom practised by Scots that required a dark-haired man to be the first over the threshold after midnight to bring in the new year.

Daisy and his young wife Ethel were planning to dance at the Pygmy Ballroom on its marvellous sprung floor before midnight and then join their fellow Scots on the rounds of their neighbours. But Daisy had to work the evening shift in Protection mine. He debated whether to skip work, but decided he was lucky to have a job, and had better not risk losing it, especially with a ten-month-old son to support.

As Ethel laid out his dancing clothes for later, Daisy debated whether he would wear rubber boots or hobnails for work. Rubber boots made sense when the floor of the mine was wet, but hobnails were more comfortable. Twice he put on the rubber boots only to take them off and finally choose the hobnails. With a promise to meet Ethel at the Pygmy in time to go first footing, he went to work the afternoon shift in the Cobble Hill section of Number One mine.

> There was going to be a great time in Nanaimo that night and some of the boys were rushin' out to get to the dance a little early before shift change. You could see their lights going by as they went along the main motor level. So we decided to pack up and go too. We could see this light wavin' ahead of us and all of a sudden – whoof. I says, "We'd better go and have a look." It was poor Daisy and he was dead. God knows what happened. The only thing we could see was a burn mark on his temple.

Daisy had been walking along the main motor level with a friend when he noticed some machinery that intrigued him sitting on a siding. As he stood on a rail to get a better look, his head touched a copper trolley wire and he died instantly of electric shock. Had he been wearing his rubber boots, he would have been all right.

When the news reached Nanaimo, orchestras fell silent in one dance hall after another. Grieving family and friends trudged home through the heavy snow that had fallen that night. Four days later, players from the city football team carried their teammate to the cemetery.

The main motor level where Daisy died acted like a reverse umbilical cord, the offspring feeding the parent, between Protection and Number One mines. On heavy steel rails, the little electric engines fed by trolley wires strung along the roof worked the main level, pulling as many as seventy loaded cars to the scales at the foot of Number One shaft. With only 1.3 metres of headroom in the tunnel, miners walking hunched over regularly touched the wire with their heads or their picks.

Ideally, a man should not have found it necessary to walk through on the main motor level. The motor transported coal and empty cars while the miners made the trip across the harbour on a scow or barge named the *Rainbow*, pushed and pulled by tugboats. The *Alert* was one of the earlier tugs used for this purpose; later it was the *We Two*. But a man who slept in would miss the tug and have no alternative but to run through the motor level, being knocked down two or three times when he touched the wire and praying that he would find a widened place in the rib to duck into when the motor sped by.

John Pecnik was a mule driver on night shift on Protection when his car loaded with rock came off the track and pinned him in a low spot. There was so little room to manoeuvre that he was stuck there for three or four hours until the day shift came on and freed him. Then mule and man took off for the Protection shaft bottom so John could catch the *We Two* and get home before his wife started worrying about him.

> There were narrow spots and then sidings every coupla hundred yards and the first siding I got to I left the car of rock there and the mule took off. Never seen it no more. The first train of the morning was due to come in with 130, 140 empty cars.

With the mule heading for the Protection stable and likely to meet the motor coming in, John was worried about a fatal collision.

> So I'm runnin' head down so I wouldn't hit the trolley and every once in a while I stop to listen for the motor comin'. I'm keeping an eye out for places where I can get off the rails if I have to. So I didn't see a dead mule so I kept goin' and pretty soon I could hear the motor, but I couldn't see it. That motor's got big lights on it just like a locomotive, but the tunnel's all twisty. Finally I find a wide spot and the motor goes by and the engineer shakes his fist at me. I get to the stables about five or ten minutes behind the mule. The boss sees me and hollers and I gets mad and I called him everything I could think of and quit.

There was no one at shaft bottom to operate the cage so John walked all the way back under the harbour along the low motor level with its threatening wires. He came up the Number One cage on top of a load of coal.

> We went up so fast I thought that car of coal was coming up through my stomach. So I got to the top at 11:00 a.m. and then I had to walk all the way from Number One to where my bike was at the *We Two* wharf. My wife was pretty worried. Supposed to be home around 7:30.

Every working day morning at 6:35, Captain Dan Martin of the tug *We Two* blew a warning whistle, its two tones echoing the tug's name,

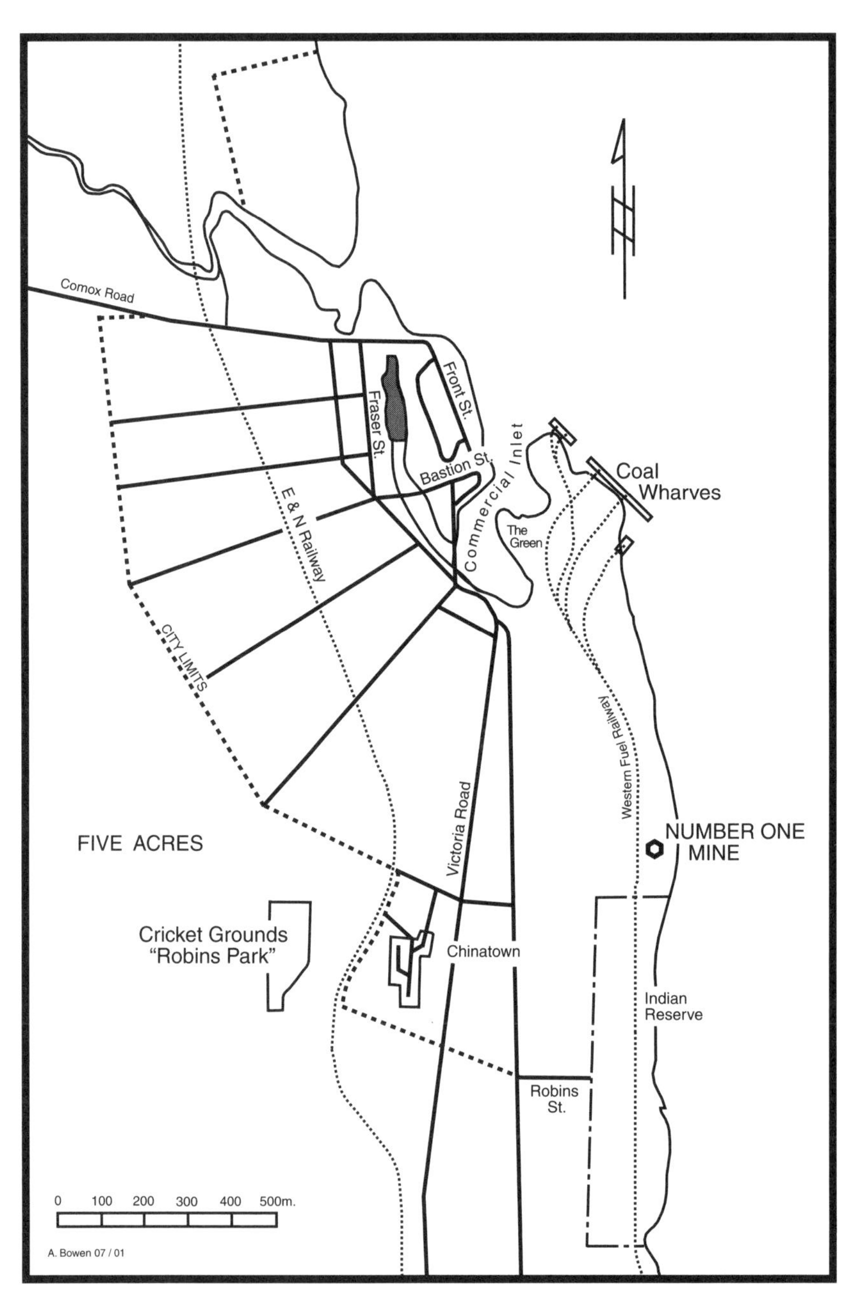

The City of Nanaimo, circa 1900, showing Commercial Inlet, Number One Mine, Chinatown, and the Cricket Grounds.

telling the Protection day shift to shake a leg. The tug would leave in five minutes. In the winter when it was still dark at 6:35, an early-rising observer could see the men running to catch the scow, the hobnails on their boots striking sparks on the road surface as they raced downhill. The tug, pulling its red-painted scow *Rainbow* and its cargo of up to two hundred miners, left promptly at 6:40. In winter a cover protected the passengers from icy winds during the three-kilometre voyage, but in summer the barge was open.

After delivering its human cargo, the tug spent the day pulling scows loaded with waste rock from the mine. Through doors in the bottom, which swung open when the crew knocked out a peg, the rock fell into the outer harbour. The *We Two* also brought steam coal from Number One for the Protection boiler in standard gauge four-wheeled coal cars, which had come from England on the sailing ships in the very early days of coal mining. All that is left of them now is one axle and two wheels mounted on a post in a downtown Nanaimo park.

The *We Two* ended her days in Victoria pulling garbage scows out into Juan de Fuca Strait, but the clandestine newspaper, *We Too*, which was printed by the union in the 1930s borrowed her name. That paper, printed, and distributed in secrecy, revealed the reality of travel on the scow.

> First of all there's the scow which is poorly heated and so dark inside that we fall over our fellow slaves when trying to find a seat on dull days. When we get halfway over, there is a general rush for the door, and also through the windows. This is done in order to get up to the front and be first off the other side. Then comes the 100-yard dash up to the lamp cabin and so to the shaft. If you don't get down the shaft soon enough there will probably be a bawling out waiting for you. And so starts the day's work.

But when it came to going home at the end of the shift, it was another story.

> Of the appalling conditions in Protection and Number One perhaps one of the major grievances is the *We Two*. It is noticeable when the scow is preparing to leave for Protection to take the miners on shift that no time is lost. We all know the mad haste which has to be made in the interval between docking and going down the mine. What happens though at the end of the shift when we arrive on top in our sweaty clothes? Do we find the bosses in a hurry to let us get away? We do not. We can sit around until anything up to forty minutes after quitting time, cold and miserable.

The editor of the *We Too* invented a slogan which he hoped would focus his readers' attention on just one of their grievances. "SCOW TO LEAVE PROTECTION SIDE AT TWENTY MINUTES AFTER THE HOUR."

When it finally departed from the Protection side, the *We Two* carried a wet and dirty cargo of tired miners to the CPR wharf. As soon as the barge bumped wood, however, the entire shift leapt to the dock and bounded up the hill, their lunch pails swinging, their feet carrying them home as fast as they could go.

Inside those lunch buckets would still be some crusts from a sandwich, a little tea or water, and sometimes even a bite of cookie. Miners' children greeted their fathers knowing that their dads saved that bite of cookie for the first of their children to greet them. But there was a more somber reason for saving the remnants of lunch.

In the large metal bucket constructed like a double boiler, the miner's wife would have placed water or tea in the bottom and sandwiches and sweets in the top. As he descended into the mine he carried with him not only his lunch but an emergency meal as well. When miners gulped their lunches between loading coal into cars, they always saved something to drink and the crust of the sandwich made black by their coal-impregnated fingers. No one would eat blackened bread under normal circumstances, but a starving man trapped by a cave-in would not be so particular. If nothing else, the saved water and crusts were talismans against disaster.

The mine is an overwhelming influence in any coal mining town. It has the power to sustain life or end it. Nanaimo feels the mines' influence in extra measure because much of the downtown sits on old mine workings from the 1850s and the rest of downtown sits on rock taken from Number One.

Nineteenth-century Nanaimo resembled a fat peninsula joined to Vancouver Island by a rocky strip where the Comox Road traffic lights are in modern times. At low tide the ravine that separated the peninsula from the island was too mucky to walk or drive across and too smelly for any civilized person to tolerate. The situation required the building of three bridges: the Bastion Street Bridge, which joined Fitzwilliam and Bastion Streets, the Long Bridge, which connected Victoria Crescent to Commercial Street and was for vehicles only, and a footbridge, which pub owners on Victoria Crescent built for those of their customers who preferred to walk. Just east of the Long Bridge the ravine widened out into Commercial Inlet. The inner end of the inlet also dried at low tide, but the outer reaches provided moorage for small boats and floathouses. On the other side of Commercial Inlet, the coal company loading dock jutted out into the harbour on Cameron Island where the Chinese who worked on the wharfs lived in bunkhouses and raised chickens close by. The site was an island in name only, a connection between it and a recreation area closer to the town having been made in the 1860s. The Green, or, as it was later known, the Central Sports Ground, was where the community met for picnics and sporting events in the shadow of the loading dock trestles.

The filling of the ravine and Commercial Inlet began in 1895 as a joint venture between the town and the coal company. Gradually, waste rock from the mines filled the ravine; waste water from abandoned mines and rain water from storm sewers drained through a wooden flume buried in the eight metres of conglomerate rock that formed the roadbed for Terminal Avenue, now the main artery through the city. The fill reached, and continued into, Commercial Inlet in 1900. By 1960, the stores on Commercial Street, whose back walls had been supported on wooden pilings, were firmly seated in rock, the small boats and floathouses had lost their moorage, and Nanaimo had new and valuable land on which a community theatre would rise in the 1990s.

The mine rock that filled Commercial Inlet presented a challenge for the theatre builders in this earthquake-conscious age, and it had been a problem many years before because it contained enough residual coal to make it flammable. Like a slow-burning fuse, the coal under pressure had smoldered.

> That mine rock caught fire and it burnt there for years. It also burned the wooden flume and that flume collapsed and it sealed off the drainage water including the water from an old mine that came all the way down Nicol Street from the Gordon estate. So that water gradually accumulated until it formed a lake at the end of Victoria Road. Now the water was under terrific pressure and it found a thin spot in a building behind the old fire hall. So they left the fire hall doors open and the water run in the back door and out the front door and it ran down Victoria Crescent and eventually found its way into the saltchuck. They had to dig down through all that rock to put in a new cement pipe and that kept a few men working during the Depression.

By the time of the Great Depression, the enclave where Nanaimo's Chinese lived had moved away from downtown. They had been banned from working underground for over forty years but they still provided an exotic presence. At the end of shifts a long line of pigtailed Chinese men walked single file from Number One, over the ridge, and across Victoria Road to a walled and gated wooden ghetto on Pine and Hecate Streets. White people made jokes about this style of walking, likening the Chinese to ducks waddling in a line behind their mother. As with most things, however, the Chinese had a good reason for what they did, a reason that went back to their approach to life and to the environment from which they came.

> In the olden days you very seldom saw two Chinese walking together, side by side. At that time the white children would say, "Single file, monkey

style." The Chinese were doing that for a purpose. You respect your elder so you let your elder walk first, that's one. The other reason is in China all the rice paddy is in the water. And they build a kind of enclosure like a bank all around and it is only so wide. So two people can't walk together so therefore they follow each other.

The traditions of China remained strong for the Chinese immigrant because many of them planned to stay only temporarily. Once they had put aside enough money they had every intention of returning to China to buy land or retire. Restrictive immigration laws had allowed only a few Chinese women to come to Canada. Either a Chinese man married in China and left his wife behind or he resigned himself to a permanent single state. Either way, it meant that most Chinese men living in the coal towns of Vancouver Island were bachelors.

Lots of people came here from Europe, from England, and Scotland to make a better life. But what are the chances of a happier life when you are cut off from your family? After they batch for a while that's all gone. They forget about it. At that time, none of the Chinese even if they are born here, even if they are naturalized, they wouldn't have a chance to vote in an election. Not 'til after the Second World War.

Most of the wooden buildings that lined the single street of Nanaimo's Chinatown were stores in front with rooms behind and above. In each of these rooms, several Chinese men lived together, pooling their expenses to increase the amount left over from their meagre pay.

Instead of cook for themselves, some of them get together maybe ten or twelve and they hire a cook. Maybe once a week the cook buy all the special ingredients, the seasoning from the Chinese merchants. And the men get paid once a month but the money goes to the contractor first and he go to the merchant and pay off everybody's bill and he take off room and board and what's left he give to the men. So that's business.

One Chinese was the leader and he contracted for the labour. And he was paid so much for a certain amount of labour being done. And he hired the Chinese. So that he was really the boss and the paymaster. Of course, he was the one who made the profit. This was one of the early methods of getting labour and I suppose a very simple way, for most companies. I don't think it was too fair for the individual Chinese, though. Because it was indentured labour, really. Mind you, these contractors probably paid their way out from China and they were repaying, but...

The "bossy man" ran a real racket.

The contractor was "above" coal mining. He was a gentleman Chinaman if you want to put it that way. He didn't believe in dirtying his hands. He would bring the Chinamen down to the doctor's office and talk for them and tell what was wrong and he got paid by them for doing it.

Groups of Chinese labourers were a common sight. The gang loading at the coal wharf was Chinese, the men repairing the mine railroad grade were Chinese, so were the firemen stoking the boilers to produce steam to run the mine machinery, and most of the crew on the picking tables. They were everywhere but in the mines.

No longer an economic threat since they could not hold the higher paying jobs in the mine, the Chinese were the objects of amused tolerance and mild curiosity. People said they loved to gamble and it was true. Fan-tan and card games filled their evenings. People said they smoked opium and that was true too. And the Chinese welcomed white people to their colourful and festive Chinese New Year's Eve celebrations. To men cut off from their own sons and daughters, the white children were especially welcome.

The Pine and Hecate Chinatown was the third in Nanaimo's history. One thousand people lived on the five-hectare site and many more gathered there.

> On Saturday night the population of Chinatown more than doubled. There was quite a few Chinese at Extension, Ladysmith, and all the other districts and they all come to Nanaimo on the weekend. They come up on the 7:30 E & N Saturday evening and there were so many get off at Chinatown that the train make a special stop right on the track at Pine Street. And then the same thing happen the next afternoon at 2:30. They would stop and pick them up again.

Chinatown was their refuge, its homely wooden buildings sheltering their traditions and supporting them in their dream to return to China rich men. The dream became more and more unlikely as the mines closed. Canada finally allowed Chinese women and unmarried children to enter legally again after 1947. A new dream replaced the old one. The future of the Chinese was in this country, and they became westernized. Their tradition of working hard made them prosperous, and their tradition of family was nurtured in a network of cousins, aunties and grandchildren who lived nearby, not far away in China.

The need for refuge died. Families moved away from Chinatown. The old bachelors stayed behind, too old to participate in the new dream. Fire hastened the transformation in 1960 when it consumed every building. The old bachelors moved to special homes built with the insurance money, their sacrifices as young men meaningless in a modern world they did not understand.

The last days of Number One came during the Great Depression in the 1930s. The mine closed but not because the coal had run out. The miners say there is more coal left than was ever taken out. And there was no labour shortage. Lines of fifty to one hundred men waited outside the

Western Fuel office every day willing to work. It was the markets that ran out. Fuel oil was taking over from coal.

For a few years the company stockpiled coal on the old Sports Grounds against some future order. When an order came, a man could get work loading it into 4.5-tonne cars. Each man got about three days' work a month. There was still some mining done but often things were closed down for three or four days at a time. When that happened they brought the horses and mules back to the main stable. Then the mine was deserted except for the man sent down to make repairs.

> I looked after the motor in Number One just before it closed. So one day the boss phones me to go and replace a rail and I had to go down alone, which should never happen. So there were smaller barns closer to the workings but they were empty and I was workin' near one of them. Then I decided to sit down and eat. In the mines there's lots of rats. The ground was littered with rats. The horses had been away so there was nothing for the rats to eat. They could smell my lunch, my open bucket. Boy I'm pickin' up chunks of rock and coal and throwin' at 'em and they're climbin' over me. I'm telling you I got scared. I just closed my bucket and threw it in the car and I started workin' as fast as I could to get the job finished and get out of there.

Soon even the rats would leave. On October 5, 1938, ten men on the afternoon shift were extracting pillars. They had just loaded twenty cars and were waiting for the shift to end when there was a cave-in one hundred metres farther in. They got eight more cars of good coal from that cave-in. The shift ended. The cage lifted the miners and the coal upward and the mine closed.

The oldest operating coal mine in B.C. produced generously to the end. Nanaimo miners had dug 16,400,000 tonnes from beneath the harbour. It took eight months to remove the winches, pipes, and rails for salvage or use in other mines. For another eleven years the headgear remained as if the town was reluctant to part with the familiar silhouette on the edge of the harbour. Then on March 1, 1950, in a few seconds, mining came to an abrupt end when the structure was pulled down.

Chapter VI

The Union Makes a Stand

The emotional high point in the life of any twentieth-century Vancouver Island coal miner was the strike of 1912 to 1914. Whether he lived through the Big Strike himself or grew up on Big Strike stories, whether his roots were British socialist, Chinese labourer or Italian immigrant, whether he lived in Cumberland or Ladysmith, was a striker or a strike-breaker, the Big Strike affected his life deeply.

Four players were involved: the coal companies, the union, the government, and the miners. Each had its own goals, and each used its own weapons to further these goals. There were four coal companies, two big and two small. The biggest, Canadian Collieries (Dunsmuir) Ltd., owned mines in Cumberland and Extension, and in 1912 was a newly formed company, having purchased all the Dunsmuir coal holdings two years before for $11,000,000. The retention of "Dunsmuir" in the company name branded Canadian Collieries, fairly or unfairly, with the reputation of its predecessor.

Also newly formed in the previous decade, Western Fuel owned Nanaimo's Number One, Protection, and Brechin mines. This company had inherited a reputation of an entirely different sort when it purchased the New Vancouver Coal Mining and Land Company (Nanaimo Coal Company) in 1903. Under its previous manager, Sam Robins, the company had co-operated with fledgling unions and had a reputation as an enlightened employer. The new owner, however, was hostile to unions.

The two smaller companies were Pacific Coast Collieries (PCC), owner of the Fiddick mine in South Wellington, and the Vancouver and Nanaimo Coal Company, owner of the Jingle Pot mine near Nanaimo. Both these companies were swept along by the events of the strike, but each behaved differently. The Jingle Pot mine settled with the union and resumed operations almost a year before the strike ended. The PCC mine attempted to continue production by importing strikebreakers, a decision that led to rioting in the streets of South Wellington.

The four companies shared common ground in their opposition to unions, especially unions made strong by international affiliations. James Dunsmuir could have been speaking for them all when he had said, "I object to all unions, federated or local, or any other kind. I think I can treat with my own men without the interference of a union." Added to the fear of organized labour was the knowledge that fuel oil was gradually replacing coal. To maintain profits, employers had to spend as little as possible on labour; since unionized workers earned higher wages, employers had to oppose unions at all costs.

The miners had been struggling to form unions on Vancouver Island for fifty years. It had begun in the 1860s with small ineffectual strikes in the Nanaimo Coal Company mines. A strike in 1877 by Wellington miners had ended after Robert Dunsmuir called in the militia to help the sheriff evict striking miners from company houses. Several miners went to jail and strikebreakers kept the mines working. In the 1880s, the Knights of Labor, an American organization that called itself "more than a trade union" and preferred political action to strikes, provided a focus for worker frustration and spoke for the miners in public. By the end of the decade, the Miners' and Mine Labourers' Protective Association (MMLPA), a locally based union affiliated with the Trades and Labor Congress of Canada, had joined the fray. When the Wellington miners struck again in 1890, the MMLPA provided strike pay, but the union was not strong enough to oppose the Dunsmuirs, who again called in the militia. In Nanaimo, by contrast, Sam Robins co-operated with the MMLPA. For twenty years, union and management found common ground in their mutual desire to get the maximum amount of coal out of the ground in as safe a manner as possible.

By the time Western Fuel bought the Nanaimo mines in 1903, the Western Federation of Miners (WFM) had absorbed the locally based Nanaimo Miners' Union, successor to the MMLPA. But the WFM was a radical American union and was unlikely to have the same good relationship with the employer. An eleven-day strike in February 1903 yielded the union nothing despite its success in recruiting 550 of the 600 men employed underground. In March, however, the Extension miners met at the Finn Hall in Ladysmith and resolved to organize under the WFM too.

The sense of urgency that accompanied the formation of the Enterprise Union No. 181, WFM, was apparent on that first Sunday, March 15th. Nomination and election of a full slate of officers, trustees, committees, and an executive board proceeded quickly. Membership on the various committees included at least one man from each of the eleven nationalities in the workforce. William Baker, an official WFM organizer, guided men with little experience in conducting a meeting using *Robert's Rules of Order*.

Within two days, James Dunsmuir had discharged seven of the union's leaders. The afternoon shift refused to work and went instead to a mass meeting. Dunsmuir threatened to close the Extension mines for two years and concentrate on developing the Cumberland mines. But still new members poured in – six hundred in the first six days. Dunsmuir was true to his word: the mines closed and the miners removed their tools.

On April 1st, a committee of three – Oscar Mottishaw, Joseph Jeffries, and Joseph Jones – attempted to see James Dunsmuir in his Victoria office, but he refused to meet with them because they represented the WFM. Five days later, when the miners in Cumberland organized Local 156, WFM, Dunsmuir threatened to close those mines in retaliation too.

Seventeen days later, the Extension miners' committee, this time without Joseph Jones, who happened to be black, stood outside the door of Dunsmuir's office presenting a card that read "Extension Committee." Dunsmuir allowed them to enter, but once they were inside, they admitted they were from the WFM. When Dunsmuir described the meeting to the hastily assembled Royal Commission on Industrial Disputes in the Province of British Columbia two weeks later, he used the racist language of the time.

> I asked them where the nigger was. They said he had not come this time. I said I heard there were objections going around that I would not see the deputation because there was a nigger on it. I told them I did not care whether it was composed of niggers, Chinese, Japanese, Indians or white men – I would see them as long as they were my own men. I went on and told them about the union and about all these agitators who were only sucking the blood out of them, that it was better to follow me than a man like Baker. He was not giving them bread.

At the regular union meeting two days later, the membership agreed that "any funds coming to this union be divided to each member and according to his family." The union was preparing for a fight; it would suspend the $1-per-month union dues until after the men went back to work. Letters of support came in from WFM locals from all over the province and from union headquarters in Denver, Colorado.

By then, the federal government had intervened with a Royal Commission that began to hear testimony on May 4th. The Cumberland and Extension locals together hired a lawyer to guide them in their testimony. The commissioners suggested to the union members that they temporarily resign their affiliation with the WFM and go to work on whatever terms Dunsmuir offered. By a margin of 216 to 49, the members "respectfully declined" the Commission's advice; they asked union headquarters to provide $14,000 per month for relief and settled in to stare Dunsmuir down.

But the Royal Commission dealt the miners a body blow. It found that there were two types of unions, legitimate and revolutionary socialist, the WFM falling into the latter category. Such unions were to be outlawed, and in their place the commissioners recommended employee associations – groups of miners within each mine who would meet with their employer to settle grievances. This was company unionism and it would persist in Vancouver Island mines well into the 1930s and beyond in some instances. Dunsmuir liked the idea, and when a union committee came to him on July 2nd to ask for a proposition to take back to its membership, he readily agreed and even offered to discharge Chinese miners from the Extension mine as a sign of his good intentions. All the miners had to do was come back to work, agree to a company union, and accept a fixed wage.

The committee went back to the old Finn Hall in Ladysmith where three hundred men, a much-reduced number from a few weeks before, waited to hear what Dunsmuir had said. A call went out for interpreters to explain the report to the members who spoke English poorly, but those men said they already understood what the proposition said. In a secret ballot 168 men voted in favour of Dunsmuir's proposal, 117 voted against.

Less than four weeks later, at a regular meeting of the now-outlawed union, the only item on the agenda was the disbanding of the local. A motion to use any surplus money to assist any "unfortunate brother (who may have failed to obtain work) to leave in search of work elsewhere" was the final act of the local that died in its infancy.

The union movement on the Island staggered under the blow dealt by the Royal Commission. For the next few years, Vancouver Island miners directed their energies to electing socialist candidates to the provincial legislature. In 1903, voters in the provincial riding of Newcastle, which almost surrounded Nanaimo in a horseshoe comprised of Northfield, Old Wellington, South Wellington, Extension, and Ladysmith, elected a thirty-year-old miner named Parker Williams. Voters in the Nanaimo riding chose James Hurst "Big Jim" Hawthornthwaite, a former colliery clerk. Williams and Hawthornthwaite, a Welshman and an Irishman respectively, constituted the entire caucus of the Socialist Party of British Columbia (SPBC) in the first legislative session to recognize political parties.

Although sandwiched between a Liberal government and a Conservative opposition, the two men in league with the Labour Member of the Legislative Assembly (MLA) from Slocan had a disproportionate amount of influence in the legislature. Williams was a down-to-earth man who was more interested in his constituents than in party labels. Hawthornthwaite was a legislative force to be reckoned with.

His first important bill provided for an examination of competency for coal miners to prevent unskilled men from working on the face. He had induced the government to design and pass the first Workmen's

Compensation Act in Canada in 1902. When his 1905 "Eight-hours-bank-to-bank" amendment to the CMR Act passed, the legislature gallery, which was full of miners, broke into loud applause. His popularity among the miners reached even greater heights when he moved to impeach the lieutenant-governor, one James Dunsmuir, over his refusal to assent to the 1907 Natal Act. Four years later, John Place took over Hawthornthwaite's seat in the March 1912 election. Place and Williams would constitute two thirds of the entire opposition to Premier Sir Richard McBride's Conservatives during the years of the Big Strike.

In 1912, Premier McBride was the king of the mountain. Recently knighted, possessor of an overwhelming majority in the legislature, riding the crest of boom times, charming, and popular, he was anxious to maintain the wave of prosperity and enhance his association with it. As his own Minister of Mines, he found it "intolerable" that coal miners should think they had the right to make demands on mine owners. The fact that they happened to be represented by the two socialist MLAs made it especially intolerable. It was McBride's aim to solve the labour problems with as little disruption to productivity as possible.

With only two MLAs to fight for them in the legislature, the miners of Vancouver Island turned again to a union. By November 1911, they had invited the aggressive and spectacularly successful international union, the United Mine Workers of America (UMWA), to organize Island mines. The UMWA faced the same problem encountered over and over again by its predecessors when they attempted to force employers to recognize and deal with organized labour: how can a union negotiate recognition with a company that refuses to recognize the union's right to negotiate or even exist? But America's biggest union, representing over 300,000 workers, felt itself up to the task. The UMWA was prepared to ignore valid collective agreements and bypass the arbitration requirements of the 1907 Industrial Disputes Investigation (IDI) Act, designed to solve disputes between management and labour before the two sides resorted to strike or lockout.

The UMWA had to think of its members in the United States. American coal dug by unionized miners could not compete with cheaper, higher quality, non-unionized coal from B.C. The forthcoming opening of the Panama Canal posed another threat to the UMWA, which feared that the new waterway would bring a flood of cheaper European labour. Such an excess of labour would place employers in a position of insurmountable strength.

Represented by an international union that did not necessarily consider their interests over those of American miners, faced with the power and influence of four coal companies and a popular government with a strong majority, the fourth party in the strike seemed puny by comparison. It was the miners of Vancouver Island in all their disparity. Lacking resources and education, all they had was their anger generated by years of failed attempts to remedy their legitimate grievances.

> When they dump your car if they found 50 pound of rock in it you don't lose the car; if they find 51 pound you lose your car. You don't get paid for that car. Now if they find 100 pounds you don't have to see the superintendent, but if they find 101 pounds you have to go see him. And then the superintendent, it's up to him to decide if he was your friend or if he's in a good mood well he might say, "Try to do better." If not, he'll say, "Well, pick up your tools."

> Because if you haven't got a union the company can say, "You're fired today" and you can't do nothing about it, eh? At that time if you lamed an animal in the mines they didn't like it. Even if you were a good worker, if they didn't like you, out you go and there's nothing you can do about it.

> A human being was nothing. An animal, yes. The mule or the horse in the mine was worth more than a human being.

In the twenty-eight years between the opening of Number One mine in Nanaimo in 1884, through the final years of the old Wellington mines and the development of the Cumberland, Extension and South Wellington mines to the year 1912, 373 men had died in Vancouver Island mines from gas explosions alone. Since 1888, the men themselves had been paying gas committees to check the mines for gas and post their findings in conspicuous places.

> They had gas committees but lots of times the gas was ignored and guys would be passin' out, headaches and what not. They used to let that happen. Yeah, the rotten buggers. We were treated like rats.

> If you smelled gas you had every right to go up and report it but they would dock your pay to go up to the top and say, "I smelled some gas." They would sit there and say, "That's two hours work lost, that's two hours you don't get paid.

> If you reported gas, it wasn't worth your job. They had what they'd call the gas committee but very seldom did they mention gas though there was gas there. Because they were in jeopardy, because they were only working men like you and me and there was no union.

> Who wouldn't fight if there were laws for the safety of the miner and they were not enforced? How'd you like to go into the bowels of the earth if you wasn't sure at what minute an explosion might come?

The CMR Act officially sanctioned the appointment of worker-paid-and-manned gas committees. When gas was present in dangerous amounts, miners were required to withdraw from the area until it was safe to return. In Extension, scene of the most recent explosion in 1909, the men were especially aware of the importance of the gas committee. They knew that as far as the company was concerned, the gas committee interfered with production and they knew that when the committee posted notice of dangerous areas, the signs would disappear almost as soon as they went up.

Isaac Portrey and Oscar Mottishaw comprised the worker-appointed gas committee at Extension. On June 15, 1912, they found gas in five working places in Canadian Collieries Number Two Extension mine. On June 25th the Ministry of Mines in Victoria received the committee's report, and the chief inspector came to Extension to check for gas in the presence of the committee and the superintendent of the mine. When the inspector confirmed the committee's findings, he ordered the superintendent to comply with the CMR Act by using only safety lamps in that area and reducing the number of men working there.

Portrey and Mottishaw went back to work, but when Mottishaw's place ran out, though he daily applied for work and watched other miners being hired, the company was unable to find him a new place. A miner with no place to dig coal has, in effect, been fired. Mottishaw went to Cumberland to seek work.

The UMWA had a special interest in the Cumberland mines. The heavily faulted nature of Vancouver Island seams meant that all companies had to hire more than the usual number of unskilled workers or mine labourers. Because these mines had an abnormally large amount of waste rock, more labourers were needed to remove the rock from the mines and sort it from the coal on the picking tables at the pithead. No union except the now-defunct MMLPA had been interested in these men, but their ability to learn mining techniques by observation and thus replace skilled miners made them very important to the UMWA.

Canadian Collieries (Dunsmuir) Ltd., owner of the Cumberland mines, had made a special effort to become less dependent on skilled miners. The company had converted the mines to electricity and had built a power plant to supply the necessary energy to drive its modern coal-cutting machinery, which was especially well suited to the longwall method used in these mines. The mechanics and engineers required to drive this machinery had a union of their own and were not interested in the UMWA. With machines mining and transporting the coal, the mines could operate with many less skilled miners. A company operating without skilled miners could maintain full production during a strike.

Oscar Mottishaw came from a family of skilled miners, a family that had been committed to the union movement for at least two generations.

Mottishaws had dug coal in the Nanaimo mines and for the Dunsmuirs in Wellington and Extension; a Mottishaw had died in the 1901 fire in Extension mine; Mottishaws had been officers of the Enterprise Local 181 and on the committee that met with James Dunsmuir during the 1903 strike. Twenty-three-year-old Oscar Mottishaw was just the man to do the union's job.

Hired as a mule driver by a contractor in Cumberland, Mottishaw worked only long enough for Superintendent W.L. Coulson to notice that he was being paid higher wages than the mule drivers employed directly by the company. He told the contractor that Mottishaw must go. Local miners appointed a committee to discuss the situation with Coulson but he refused on two occasions to meet with them. When Mottishaw went to see him on his own behalf, Coulson said the company "reserved the right to hire and dismiss without having to answer to anyone for its decisions."

> This is what was supposed to have triggered the strike. But in the meantime, the United Mine Workers already had men organized to some extent. Otherwise they would never have walked out spontaneously which they did.

> But I think the company was looking for an excuse then. They knew the men was starting to organize. The Mine Workers was organizing them.

The UMWA had been on the Island for less than a year, but had already attracted four hundred members in Extension, two hundred in Nanaimo, and three hundred in Cumberland. When the Cumberland local called a meeting on the night of September 15, 1912, the members agreed to a decision taken earlier in the day by a mass meeting of company employees that a "holiday" be declared for the following day to protest Mottishaw's firing. On September 16th, while the miners took the holiday, the superintendent rebuffed a third committee comprised of both union and non-union men.

At seven o'clock on the morning of September 17th, notices at all the Cumberland mines greeted the men when they reported for work. Every man must take his tools out of the mine and must sign an individual two-year contract under the old conditions. The notices barred two thousand men from their jobs.

Later in the strike the union would insist that they intended the holiday to last for only one day and that the company had locked the miners out. But the company claimed the men had every intention of staying out longer and that constituted an illegal strike under the IDI Act, which required thirty days' notice before calling a strike.

Word of the Cumberland holiday and lockout reached Ladysmith and Extension. A meeting of the Extension local with an international union organizer named George Pettigrew present voted to declare a holiday in sympathy with the Cumberland miners that would last until they were allowed to return to work. The same lockout notices appeared at the entrance of the Extension tunnel.

> We held [a] meeting down in Ladysmith and 1,700 men joined the union in one day. They didn't know what they were doing; they thought they were going to be big shots. Extension and those places were always small, small villages and they didn't have enough work to keep them going. They were getting held back so they joined. They shut down the mine there but it didn't do no good. Most of the miners thought the strike was just a holiday and would be over in a short time.

Neither the union nor the company was interested in using the word "strike" in the rhetoric and both wished to avoid a clause of the IDI Act that required the warring parties to place their dispute before a board of inquiry under threat of a severe penalty – the union because it would not gain recognition through such a process and the company because the act required the employer to give thirty days notice before locking its employees out.

The company devised a charade so that it would appear as if there were no lockout. Every morning for thirty days, the mine whistles at Extension and Cumberland called the men to work. But no union man was working, and in Cumberland, the Chinese miners refused to work as well. Despite the animosity felt against them, the Chinese had been interested in the benefits of collective action for at least as far back as 1906 when they attempted to strike for higher wages.

Then on September 24th approximately twelve provincial policemen specially recruited for the job arrived in Cumberland and denied white miners access to Chinatown and Japtown. The next day the Asian workers agreed to a two-year contract with the company. Rumours flew. Some said the Chinese and Japanese had been threatened with deportation; others said two hundred Chinese had been imported to swamp the vote in favour of the company; still others reported that the company had instructed a doctor to tell the Chinese how to vote.

> Up in Cumberland you see, you had an awful bunch of Chinese workin' up there, that time. And Japanese.

> Yes, you couldn't blame the Chinaman in a way. Some of them worked for as low as a dollar a day. And that Chinatown, just a bunch o' shacks, you know. They used to say there was hundreds of Chinamen there that wasn't workin'. Who can tell?

The international constitution of the UMWA contained an anti-Oriental clause, but on the Island the union, knowing that the Chinese and Japanese miners' support was crucial, had been prepared to overlook it. When the Asian men returned to work, they lost any support that the striking miners might have given them.

> I heard about one Italian fellow he went over to the Colliery dam and there were a lot of Chinamen fishing there. He walked up to the first Chinaman and says, "Where do you work?" The Chinaman says "Number Seven" and he picked him up and threw him over. And he come to the second one and did the same thing. Finally the fellows that were with him stopped him.

The return to work of Chinese, Japanese, and some white miners brought the first violence of the strike. Initially there were just a few strikers harassing the workers on their way to the mines and the occasional woman poking a straggler with the point of her umbrella. There were failed attempts to blow up bridges and set fire to pitheads and then, as the days darkened into November, mobs of up to five hundred began to "escort" the strikebreakers home from work, singing the "Marseillaise," and filling the air with hoots and screams. The end of the day shift was especially dangerous because some of the strikers had been waiting all day in the bar.

But when the police put an end to the fun, the women came up with a subtler form of harassment. At the beginning of each shift, a procession headed by several accordionists, a fiddler, and a trombonist would play the strikebreakers to work, wait for the end of the shift, and play them back again. And there were variations on the theme: sometimes women dancing the Highland fling would lead the procession.

But more ominous than dancing, rumours of beatings and gun threats were spreading about Cumberland. The mayor asked the attorney general for aid. On November 22nd, 120 special constables, 100 on foot and 20 on horseback, joined the small contingent of provincial policemen already in Cumberland. Two weeks later 17 more horses arrived. Thirty-seven horses make an impressive show, but the policemen riding them were not in uniform. That the B.C. Provincial Police wore plain clothes would be seen in later years as a tactical error that they would rectify in 1925. A uniform gave authority and engendered respect. In 1912, the populace resented these plainclothes policemen, all the more so because most of them were specials – non-policemen sworn in especially for this job and likely to have been recruited from among the unemployed of Victoria and Vancouver.

Now the company called a meeting of 250 Italian miners, whose reputation as strikebreakers made them seem malleable. Given the choice to "sign a two-year agreement or leave Cumberland," however, they chose to leave. Not all the white miners made the same choice. While the majority of strikebreakers in Cumberland were Chinese and Japanese, some white miners returned to work as well and earned for themselves evermore the label of "scab."

> The name scab must have come from the old country. They used to call 'em blackleggers, scabbies.
>
> We called 'em black Minorcas. That was a kind of chicken.
>
> When we were kids I remember Dad saying one thing about the strike. "When I go to my grave, I'll not go with a black mark against my name." At the time we didn't think nothing of it, us kids didn't, but as we got older then it dawned on us that he meant he would never scab.
>
> My father worked nine shifts. That's all he worked and let's say it was against his religion to scab. If you're a striker and a union man you don't scab. I know one man, his wife forced him to go and scab. No woman will force me. If I wanted to work I'll work, but I ain't gonna scab. I never did in my life and never will. If my kids scabbed tomorrow they wouldn't stay in my house.
>
> No matter what they do after, the name scab is still there. If your father scabbed and you worked at the mine, another mine at a later date when you grew up, they'd say, "There's that scabby so and so's son just started." You still held the name; it didn't matter. You can't blame the son but there it is.
>
> When the strike began I was a boy of about seven. I can remember the wives of the strikebreakers being escorted to the Post Office and store by special police. In those days it was all kerosene lights and we used to buy kerosene in square cans and the women used the can to boil their clothes in on wash-day. And we used to get those cans and tie a string around our necks and go down the road behind those women and we'd yell, "You dirty scab, dirty scab" and beat those tin cans.

The men who returned to work in the early days of the strike were local men motivated by defiance or fear or nagged by a worried wife. Later, as the strike persisted, the company began to recruit miners from the prairies, the United States (U.S.) and Britain. Many of these men arrived unaware of the situation on the island. In an even later stage of the strike the companies hired whatever help they could pick up on the streets of Vancouver and Victoria. These men were like some of the specials – no better than thugs. But from the beginning and increasingly as the strike dragged on, the men showing up for work in the strike-bound mines were former strikers, men who were tempted by the promise of higher wages or promotion or men who perceived the fight as hopeless, who searched their own hearts and decided they could not sacrifice their families' welfare another day in a lost cause.

On September 28, 1912, Canadian Collieries evicted all strikers from the company houses in Cumberland and Bevan. Families loaded their belongings on wagons and moved in with relatives or pitched tents on what came to be called "Strikers' Beach" in Royston, seven kilometres away, or at Comox Lake. Eventually, many strikers built cottages for themselves but all faced that first long, wet winter in makeshift accommodation.

> One of my sisters was born at the time of the strike. We lived in a company house and if you didn't scab you had to get out of that house. So we didn't have much choice. My uncle moved us about two miles from Cumberland into a two-room shack and that's where five of us lived and the sixth one was born there. We were there for two years.

> So we came to Cumberland in May 1912. We'd just got a company house and fixed it all up and my mother papered it, scrubbed it with lye. We'd just moved in when the strike took place and they threw all the furniture out on the road. If you didn't work, you didn't live there. So anyway, we went to live with some other people at the back of the old Post Office. They owned their own house. Se we lived with them all one winter, my mother and four kids, while my father came down and got a job at the Jingle Pot.

> We moved to Bevan in 1911 and when the strike come we had to move out and Dad built a house in Royston. They got $11 a week strike pay. My mother always managed. We used to have the things we needed and I don't know how she got them. I can remember going to school and my clothes were just as good as the others whose fathers were working.

With all the union men off work in Cumberland and Extension, the organizers shifted activity to Nanaimo where only two hundred out of two thousand miners belonged to the union. At a mass meeting on October 12th at the Opera House downtown, a speaker chastised the audience. "Each man is just waiting for the other man to join. For just $2 you can be members of the UMWA."

A week later, the attendance at a second meeting strained the capacity of the Opera House, but many were still reluctant to join. The men of Nanaimo were mostly employees of Western Fuel, a company which, while not exactly enlightened, had dealt more fairly with its employees and which had inherited the good will engendered by its predecessor. Nanaimo miners, too, generally owned their own modest homes and enjoyed the amenities that a larger community offers. Their jobs were not threatened by the Chinese, who had not been allowed to work underground since the 1887 explosion. In many ways, the miners of Nanaimo had much more to lose by striking.

Seven hundred striking miners in Ladysmith and Extension had chosen to stay out until the company recognized the union, but the sooner this happened the happier they'd be. UMWA president, Robert Foster, had been urging the government to intervene. The federal Minister of Labour, T.W. Crothers, attempted to invoke the IDI Act but the aggrieved parties refused to co-operate. B.C. Premier and Minister of Mines, Sir Richard McBride, had refused to interfere in September 1912 and would refuse again in May 1913 when Foster protested the conduct of the special police. A federally appointed commission held sessions on the island in February of 1913. Its report – generally factual but management-slanted – would be completed in August 1913, just in time to see a year's tension erupt in a week of rioting.

When the legislature had convened in January 1913 the two Nanaimo MLAs asked for an investigation of the strike. Their motion was defeated thirty-five votes to two. The February 12, 1913 edition of the *Vancouver Morning Sun* quoted Parker Williams' report to his constituents on his fruitless attempt to get the government to intercede.

> I talked for forty-five minutes. I tried and I succeeded to keep all party feeling out of my speech. I pleaded with them for an investigation of this strike, or walkout, call it what you will. I explained this was not a fight between capital and labour. Neither Place nor I would ask interference if it was. This is a fight against the Minister of Mines, that is the man we are striking against.

Williams had hoped that the MLAs from Vancouver would support the Opposition because the strike was causing a coal shortage in that city. But Premier McBride headed off the mutiny of his Vancouver members by promising a commission to investigate the price of coal. Williams went on in his eloquent style.

> This is a fight not primarily for a living but for the right to live. This is a fight against the undertaker and the morgue and the department of mines more than against the Canadian Collieries.
>
> Ladysmith is very quiet. The streets are deserted, the men are subdued, holding on, following meekly, turning the other cheek, obeying implicitly the suggestions of the quiet young fellow who is their leader.

The quiet young man was Sam Guthrie, a well-educated Scot who had come to Ladysmith in 1911. A future MLA himself, Guthrie was an attractive character with his flaming red hair and his new Finnish bride, Lempi, by his side. When the Extension mines resumed limited operation in April and May, the strikers and their families in Ladysmith held on, doing without, making do, hoping it would all end, finding new ways to provide for themselves, and remembering why they were striking. An anonymous writer in the same issue of the *Vancouver Morning Sun* summed it up eloquently.

> You don't forget when you see thirty graves all new dug in a row waiting to be filled with men you've known all your life. Yes, I remember when we had a funeral here every Sunday. But wouldn't you fight and starve if need be if when your man left the house you didn't know how he was coming back? Way back in your mind while you were tending and mending to keep your

children tidy you couldn't tell whether their father was coming home on a shutter dead or with his back broken by an accident.

Miners who were children during the strike are eloquent too when they remember how their mothers made do.

My mother had things tough. But what we did, we just put our shoulder to the wheel and carried on. We had an old rickety wheelbarrow with one wheel. And there was peat down at the football field, oh about 300 yards away. And we hauled hundreds of loads of that peat out 'cause it was rock where we lived. And we made a wall and built it up and we made a garden. My mother bought a pig and the neighbours had chickens. So we got a dozen eggs and the neighbour lent her clucking hen. That's what you can do when you have to. Yes, she was a good manager and very resourceful.

So it kind of come hard times and Mother had to sell the furniture. And them old country miners they pride on their furniture, nice furniture, and one of the miners when he come home said, "They must be pretty poor, the missus is sellin' a nice big sideboard."

Reasoning that it was important to keep workers from drifting away into logging, farming or railway jobs, the union paid relief to anyone who applied for it, union member or not. At first the miners received cash but many spent it in the local saloons, so the union arranged with local businesses to give credit to the amount each miner was allotted.

I was two years on miners' relief. We got $4 a man per week and $2 for a woman and $1 for a baby. I was getting $7.

A man and his wife got $6 a month and what the sam hell, I got $4 for myself and $1 for my mother. I was supporting her and they paid me the same as if she was a kid.

I don't know how we did it. It was a struggle, but he never went to work anyway. He stuck it out with the rest. They were getting' a bit of relief from the miners around Montana because they were collectin' see. They sent millions of dollars up here you know.

We got put out of the house in Cumberland and we moved to Minto. I was about five years old, but I stole to get food. You were a grown-up boy at five years. Some people would say what kind of boy was that. If you're hungry you'd do anything.

We lived on mush and beans for two and a half years.

As I recall, during the strike we lived on beans and rice and those kinds of things. But beans was a fairly staple diet, dried beans in those days. During the strike a lot of them went out and shot their own meat. They were all hunters.

Mother backed Dad up to the limit on his stand to go on strike and never once complained although things were tough. We had the best country in the world, good gardens, lots of fish, deer (we called it government veal), all kinds of shellfish and no pollution.

We all had gardens them days. That year we set a lot of potatoes. That was lovely soil. Black soil. Gee, it used to grow lovely potatoes. They were dandies. And there was a family of ten lived on the corner down here. So the old man he says, "Load up that wagon with potatoes, take them over to the neighbours."

I went through the strike. I picked coal on the slack heaps between Nanaimo and Reserve mine at 5:00 in the morning to make sure that we had heat in the house.

In fact, many a young man who had been too busy working in the mines to have time for courting now had an abundance of it and acourting they went, which resulted in an epidemic of marriages. Many of the young people like my husband and I believed in the old saying that two can live as cheap as one. Whoever fabricated such a misconception should have had their heads examined.

There was lots of things you had to go without but I don't know. Lots went out huntin' and fishin' and we didn't bother about clothes. You didn't have to worry about dressin'. A pair of overalls, a shirt and that was it. The stores in Cumberland, Simon Leiser, I think they carried me up to pretty near $70 before I was cut off. Oh eventually most of the stores cut people off but they carried people through quite a lot.

There was a woman whose husband had been killed in the explosion in Number One mine and the men set her up in the grocery business in Nanaimo on Haliburton Street. When the strike came she carried people for their groceries. There was another man in South Wellington did the same.

There had been signs of trouble in Ladysmith in the first weeks of 1913. Strikers were threatening strikebreakers as they made their way through town to the miners' train bound for Extension. On January 28th, fifty-seven miners requested protection. In response to Mayor Hillier's telegram, "Situation out of control of city police. Suggest provincial police take over city and maintain order," Attorney General Bowser sent Provincial Constable George Hannay and seven specials. The combination of the police and a warning from the UMWA that the international would withdraw support if the union men in Ladysmith did not obey the law, brought calm to the coal camps south of Nanaimo.

The strikers and their families in Cumberland, Ladysmith, and Extension emerged from the first winter of inactivity and making do bolstered by weekly union meetings. They were determined to follow the union's lead, and were aware that Canadian Collieries had been able to maintain production in Cumberland despite the strike. In keeping with its "competitive equality" principle and anxious to stop the flow of higher quality, cheaper coal from Vancouver Island, the International Executive Board of the UWMA had decided to widen the strike to all the companies on the Island. Preparations began in secret. The main target was to be the Western Fuel Company, which was about to open its new Reserve Mine amid much publicity over the large number of men it would require to dig the four-metre seam.

Robert Foster, President of District No. 28, UMWA, received a letter dated April 30, 1913 from international representative, Frank Farrington, ordering Foster to call a strike of all miners in the Nanaimo, South Wellington, and Jingle Pot mines. Included in his list of instructions was the following:

> You should also exert every effort to prevent unlawful or abusive tactics by the men during this contest, and you will also make a diligent effort to secure the names of all men who refuse to respond to the call to strike so they may be published throughout Canada, Great Britain, and the United States.

Then he issued the rallying cry: "The time is now here for the men of Nanaimo and South Wellington to prove their worth." The time for secrecy had ended. Posters declaring a strike and written in English and several other languages, including Chinese and Japanese, appeared all over town.

The next morning was May Day, a day when the children of Nanaimo could watch the parade, participate in the sports events, win prizes, and eat special food. But their parents looked forward to the speeches. Parker Williams and John Place brought greetings to the assembled crowd, then President Foster took centre stage and reminded the miners of the pamphlets they had received as they left work announcing a mass meeting at the Princess Theatre for later that evening.

Of the twelve to fifteen hundred men that packed the theatre on Selby Street that night, only a third were union members. Though the others came to be persuaded, there were many with doubts, especially when Foster told them that there would be no new strike vote: the results of the vote that had been taken six months before at the second Vancouver Island UMWA convention still stood. The union leaders had proposed a wage scale to the owners at that time and had threatened a strike. Six months had passed with no reply. Believing there was no fight in the union, Western Fuel had treated the union proposals with contempt.

In the Princess Theatre that May 1st evening the platform speakers urged the Nanaimo men to join the fight. "The men of Ladysmith and Cumberland laid down their tools to protect two men and to enforce the law. Why shouldn't Nanaimo throw in its weight?" Cries of "hear, hear" mingled with loud voices of dissent. But demands for another strike vote did not sway union officials. Over the noisy auditorium the speaker's voice thundered, "The man who goes to work tomorrow is a scab. We'll know tomorrow whether the men of Nanaimo are scabs or white men. The union office is open to initiate members," the speaker went on. "No need to worry about the membership fee until after the strike is over."

Fortunately for the undecided, the manager of Western Fuel, Thomas Stockett, had decreed that the mines would close for two days to allow the men to vote. The following evening in the same theatre an overflow crowd listened to the spokesman for the joint committees, those creatures of the 1903 Royal Commission. The speaker reminded the audience that they had signed collective agreements with the Nanaimo, South Wellington, and Jingle Pot mines and these agreements still had time to run. The UMWA only represented a small minority, he said, and had not allowed a strike vote the night before.

But the tide had turned. The audience heckled the platform party. Two thousand men filled the streets leading down to the courthouse to cast their ballots, the majority in favour of the strike.

As day followed day, more and more miners retrieved their tools from the mines. By the next union meeting, one week later, eighteen hundred men were members of the union. The fact that the union leaders had turned their salaries over to the strike fund and would be living on strike relief like everyone else had won some of them over. Parker Williams and John Place endorsed the union. The two popular MLAs had grown in stature for their lonely stand in Victoria.

By May 8th, the PCC mine had no men left and the Western Fuel mines were completely shut down. The men of the Jingle Pot mine worked one day and then walked out en masse. The union would allow fire bosses and pump men to keep the mines in repair, but this concession lasted only until strikebreakers began to work.

Up in Cumberland, Canadian Collieries had sent company recruiters to Durham to advertise for miners, married men preferred. Having been assured that the strike was over and they "could come without any fear of being scabs," sixty-two miners had sold their homes, left their wives and children to await their summons, and travelled by boat to Portland, Maine, where they had boarded a train for Montreal and points west.

The union had posted men across Canada and had distributed notices warning the "Workers of the World" to avoid Vancouver Island if they did not wish to "scab." But the company had hired guards to ride on the train with the miners to prevent them from hearing this news. It wasn't until they reached Winnipeg that the first whisper of a strike reached them. The guards dismissed it as "all lies" and the men resumed their journey, living, as one of them described it, "like fighting cocks" provided with all the food, beer, and tobacco they wanted.

In the mountain town of Revelstoke, B.C. their cushioned treatment came to an end. An official of the UMWA had managed to get a message to them. "The strike is still on." The miners confronted a company official who threatened to put them off in the bush. For the past day, the windows of the train had displayed the rugged Canadian Rockies – challenging country for a stranger. After a hurried conference, the Scottish miners pretended to go along with the company until the train reached Harrison Mills at the entrance to the Fraser Valley. Union men travelling as passengers on the train assured them they were close enough to Vancouver to reveal their true intentions.

Twenty-two Durham miners announced, "No matter what happens, we will not go to Cumberland." They left the train with the union men and told their story to the Vancouver press. The remaining forty continued on to the Island, fifteen of them on the *S.S. Charmer*, which worked the Vancouver-Nanaimo-Comox route. While the *Charmer* rested at its Nanaimo berth, unionists on the dock shouted insults at the Durham men and ten other strikebreakers bound for Cumberland.

> They went over to England and brought in some men. We got some to quit and some went to work. You couldn't blame them in a way I guess. They were told they were coming out here to work, they came out and I guess they had nothing. You know, you could meet them coming off the trains and offer them strike pay but that wasn't very much encouragement to them, coming out from the Old Country to make a fortune.
>
> They were told: "Don't go to Cumberland, there's a big strike on." But a man's tied up; he's got his family. Of course, that's an excuse; some of them didn't come. Some of them have said to me that they felt they had to come because they were obligated, their fare had been paid, and all that.

By June 1913 outsiders – twenty-three Italians from the United States – were working in the Extension mines. Then 250 men from Victoria accepted a cash advance, boarded the E & N and, although some of them jumped off the train before they reached Extension, the rest arrived to a town full of dedicated union men. To protect the strikebreakers, Canadian Collieries housed them in company houses that sat on company property – an area that came to be called the bullpen.

Another bullpen took shape at the PCC mine in South Wellington. This company had lost no time since the strike call and had resumed operations with whatever help it could hire after only one month. By July 5th the mine had twelve diggers working and fifty waiting for accommodation. The single men slept in bunkhouses, the married ones in company houses just vacated by strikers' families.

Even as the warm days of summer saw the Extension and South Wellington mines resume limited operations, the strikers' discipline had held. Quietly, they had vacated the company houses in South Wellington and quietly they had watched as outsiders took over their jobs. Part of their compliance was due to the presence of special police; part to the union's determination to obey the law; and part to its efforts to make life as tolerable as possible.

Weekly union meetings gave the men a chance to plan strategy and the leaders a chance to reinforce solidarity. The women were an important factor in the union's plan.

> The women was with the men. They used to have a dance every week. Up in Cumberland the union built a hall during the strike and they used to have a dance or perhaps three dances a week. We had a piano player and a fiddler, two strikers. The women would bring sandwiches. They didn't worry much about clothes, you went with what you had. There was nobody trying to beat the other with dress.

> The women did a lot. Hard times dances, 5 cents and quite a few of them played instruments. They had good times. In fact some of the best times you ever had in your life and you didn't realize it.

> We had a seven-piece orchestra and they would play the two-step 'til dawn. Once a week, month after month, the UMWA Ladies Auxiliary had a concert and dance in Nanaimo. Up in the Oddfellows Hall. At first they used to have imported talent, even. You'd hear the singers, then there was a potluck supper and then you'd dance. They raised money and everyone had a good time.

But outside the dance halls and in the streets, the continued presence of the special police was proving to be more of an irritant than an agency for the preservation of the peace.

> One of these special policemen would come along, you might be walking on the sidewalk, and he'd bump you off, stagger into you. You'd retaliate and you were pinched for assaulting a police officer.

> The specials belonged to the type that was lacking in the elementary principles of manhood.

> The guys who couldn't stick it out, maybe were in a pinch with four or five children and felt they had to go back to work, would have protectors to and from the mine. The police brought them down so they wouldn't be molested on the way to work.

There were 191 specials in Ladysmith, Cumberland, Extension, South Wellington, and Nanaimo. Under normal conditions Nanaimo's police force numbered five. But these were not normal conditions. In the summer of 1913 on Vancouver Island, 3,500 men were idle. Some had left for Australia or returned to England. Those who stayed concentrated on keeping the peace and keeping bread on the table.

The sympathies of the churches were divided. The Baptist minister, Reverend J.H. Howe, in a fiery sermon, declared himself a trade unionist and marched in the first strikers' parade. He accused other churches of ranging themselves "limpet-like" on the side of the capitalists. An Anglican clergyman in Victoria devoted a sermon to the strike and managed in that one sermon to come down firmly on both sides of the question. Methodists, by contrast, were sympathetic to the strikers and asked the government to intervene. One of them, Reverend J.W. Hedley, wrote one of the few contemporary accounts of the strike to survive to the present day. But if the men of the cloth were generally sympathetic, the newspapers were not.

> In them days the papers was all against the strikers.

The *Islander* in Cumberland and the *Nanaimo Herald* were most assuredly anti-strike. The *Nanaimo Free Press* made a valiant try at impartiality and in the flowery phrases used in journalism in those days gave a less partisan version of events. The Vancouver and Victoria papers, observing the events from afar, published rumours and sensationalized events.

As newspapers and religious groups took sides and as special policemen patrolled the streets and approaches to the mines, the impartial summer sun bathed all the Island coal camps. With school out, the children of union members joined the UMWA Ladies Auxiliary in a demonstration of solidarity. Little girls in white dresses with red hair ribbons and badges led a parade from the union headquarters on Nanaimo's Commercial Street. Boys sporting red badges marched to the music of the Silver Cornet Band and the Ladysmith Miners' band. The parade route led to the Cricket Grounds, where the children competed in track events for prizes and the adults listened to stirring, morale-building speeches. No one watching the happy summer scene would have predicted the ugliness and burning that would alter the mood of that summer in just over a month's time. There was even a rumour of a settlement, but the *Free Press* cautioned its readers,

> Not until it hears with its own ears the raucous voice of Number One sending out the call to labour over the hillside, and not until with its own eyes it sees the idlers taken from the streets and transformed into miners once more with coats over their arms, dinner pails swinging and cuddy pipes perfuming the sea wind as it blows over the black mouth of the pits, will Nanaimo regain her composure. But when she does, she will take a long and slow revenge upon the men who precipitated the strike. They will walk the path of the evil-doer and get no sympathy.

Chapter VII

Hurray, Hurray, We'll Drive the Scabs Away

> The law in Ladysmith today is regarded as a valued piece of Dresden china. All care must be taken not to break it. – Parker Williams, MLA

> The fight has not been a spectacular one. There have been no riots, no conflicts between pickets and strikebreakers. – *Nanaimo Free Press*, May 31, 1913

> In August of 1913, there was a riot and if ever there was a riot it was that one; it was a dandy, if I tell you myself. – George Edwards

> I was seven years old in August of 1913. I started going downtown and a man caught up with me and he seen who I was and wanted to know where I was going and I told him, "I'm going down to see where all the fun is." – Eino Kotilla

> You could hardly hear your own ear from the hollering, "Drive the scabs away." – James White

> They had a heck of a splatter. – Jock Craig

> The Ladysmith men are a distinctly lower order of people and display the traits employed by the ignorant and savage. Foreigners seem to predominate. – Pinkerton Undercover Detective

As the summer of 1913 progressed towards the middle of July, the weather turned hot and with it the tempers of the inhabitants of the coal towns. Much has been said about the riots in Cumberland in July and those around Nanaimo in early August and each speaker has an idea about who was responsible for turning a carefully controlled strike into a nightmare of rioting and anarchy.

Some say that the strikers finally reached a breaking point, provoked by the increased number of strikebreakers and made irrational by hot weather and worry about their families; some say the strikebreakers themselves provoked the riots, harassed as they were at every turn and living in enclosures requiring the protection of special police.

Others see sinister motives behind the actions of the companies, the government or the union. They point to the Cumberland riots, where the special policemen appeared to provoke the strikers deliberately. This is an example, they say, of the company manufacturing an incident to justify bringing in the militia, something the government seemed most eager to do. For others, the UMWA was the archmanipulator. The strike had lasted almost a year in Cumberland and Extension; production in Cumberland was regaining normal levels; broadening the strike to include the Nanaimo and South Wellington miners, instead of forcing a settlement, had only added more men to the strike relief rolls; scab labour was arriving in increasing numbers. Desperate action was needed.

The behaviour of the police in the early days of that summer was provocative to say the least. A large crowd of Nanaimo strikers had gathered quietly outside the roller skating rink near the Princess Theatre before a regular union meeting. Twelve mounted provincial policemen rode straight into the crowd, scattered the union men, and forced them to flee on foot. In Cumberland, specials staggered deliberately into pedestrians on the sidewalk and then arrested their victims if they retaliated. The strikers' willingness to submit peacefully to such provocation was fast disappearing.

It disappeared completely on July 16th in Cumberland when a special named Cave let it be known that he intended to clear the town of strikers. For the next three days the town watched and waited until Saturday evening, when the wooden sidewalks on either side of Dunsmuir Avenue were crowded with beer parlour patrons. When Cave and fifteen other mounted specials appeared at the end of the avenue and began to proceed down its length in single file, they had everyone's full attention. Then a striker named Reynolds took a punch at Cave who was twenty-five kilograms heavier than Reynolds and on horseback. Cave arrested Reynolds and the fight was on.

> A whole bunch of young people was standing on the street there and this one guy, feeling hot and heavy, jumped out and he smashed the first one in line. With his fist. And that started the melee.

> There was men and horses and everything going in all directions. They had clubs three feet long, pick handles, and one thing and another.

> I was on the street when it started. It was just where the fire hall is now in Cumberland. At that time there was benches along the street where the men would sit and talk. The fifteen specials were all on horseback and one of 'em pulled out a paper and unrolled it and he read something off it. I was on the corner below and all of a sudden he hollered "Charge" and the horses charged these men on the corner and the next thing I knew I got a boot in the arse and my father was behind me and said, "Get home" and I didn't argue. I headed for home. That was quite a battle and I would like to have seen the whole thing.

> I remember the riot up there during the strike. Dad came running home and he didn't have a jacket on so Mother said, "What's the matter; where's your jacket?" "There's a fella," he said, "a friend of mine up there in just his shirt sleeves and he was scared he'd get arrested because they seem to think the fighters are in the shirt sleeves so I gave him my jacket and I came running home to get another one." Dad says, "Don't let the kids out. Don't let them up there." There was lots of kids up there but we was kept in.

It was two weeks before the police arrested six of the strikers including the local union president, Joe Naylor, who had not even been on Dunsmuir Avenue that Saturday night of July 19th. And word of the events travelled just as slowly. Cumberland was two days by road from Nanaimo; the E & N and its companion telegraph lines would not reach the Comox Valley for two more years. In the time it took for the news to trickle down the Island, the story had become distorted.

> So the story got down here that the scabs had started to raise a little bit of hell with the strikers so we started to raise a little bit of hell with the scabs. That's what triggered the riots down here.

> The company got all the bums from skid road in Vancouver and Seattle to scab. In August things came to a head. The scabs got on a party and said they would clean out the union men. When this word got to Nanaimo the miners came with guns and the war started. The scabs ran into the mine in the woods and other people came in and looted and smashed homes.

There is no doubt that an increasing number of strikebreakers – local and imported – were working in the Island mines. If there had been a need on the companies' part for secrecy before, it was even more acute now that the union men were fighting back. The PCC had recruited ten men from Missouri, picked them up in Vancouver in a gasoline launch and brought them across Georgia Strait to the company docks at Boat Harbour. They climbed aboard a boxcar on a PCC coal train which took them to South Wellington where they refused to work when they realized that the strike was still on.

But it seemed as if there was an unlimited supply of strikebreakers. The Cosmopolitan Labour Exchange provided Canadian Collieries with workers throughout the strike. Some were unemployed men picked up off the streets and some were immigrants, their lack of fluency in English making them easy targets for persuasive recruiters.

> My father came up from the States to work in the mines here and he apparently come up and went to work during the strike. They were advertising for people all over. And a lot of them didn't know what they were doing and when they come up there was a lot of people working so they just went to work.

> A lot of Italians came in. Strangers. Everything was quiet and peaceable as long as the men that was working were from around here, you know, belonged to Ladysmith. But when they started to bring in these men, that really caused the trouble.

The addition of more outsiders may have seemed like a body blow to the strikers, but the union had already decided that drastic action was necessary. No one knows for sure whether the decision to organize the riots was an official UMWA decision, but union organizers, some who had been on the ground since the summer or fall of 1912, served as picket captains and muscle during the riots. Frank Farrington, from UMWA headquarters, had checked in to the Hotel Vendome in Nanaimo that summer to lead the union effort; David Irvine, a UMWA executive from Illinois, and George Pettigrew from UMWA international, attended union meetings in both Ladysmith and Cumberland at the beginning of the strike; John Walker, a former head of the UMWA from Illinois, had been seen on the Island from time to time in the previous year; American "Big Louis" Nuenthal was the president of Local 872 in South Wellington, where membership had recently jumped from ninety to 210; and Joe Angelo, the international union's "Italian organizer," was on the Island as early as October 1912.

If the Italians seemed to be singled out for special mention as strikebreakers, it was because their situation as immigrants to North America was unique. Of all western and central European countries – a major source of Canadian immigration at the time – Italy was especially poverty-stricken. Hundreds of thousands of its men had emigrated since Italy had officially become a nation state in 1870. But emigration was particularly difficult for Italian men who planned to leave their sunny country only for as long as it took to make sufficient money to enable them to buy land at home. Then they would return to Italy and live there for the rest of their lives. Leaving their families behind and promising to send money for elderly parents and dowries for marriageable sisters, Italian men – many of them unskilled – went to the rest of Europe and to North and South America. They were lonely, homesick and desperate and would work for anyone who paid them. The extent of their desperation can be measured in the fact that since at least 1880, the word "Italian" had become synonymous with "strikebreaker" in labour circles.

On August 9th on a dark road, four Italian strikebreakers attacked two Ladysmith strikers and stabbed one of them, a certain James Hatfield. The police arrested one of the Italians at Ladysmith's Temperance Hotel, a boarding house regarded as headquarters for the imported strikebreakers because so many of them lived there.

The stabbing added to the fear and hostility that festered among strikers and strikebreakers alike. The occupants of the company bullpens in

South Wellington and Extension were said to have firearms. Strikers thought the Ladysmith police were not giving them enough protection. They passed a resolution which said, "If the police do not extend to our members the protection of the law, we will be compelled to take measures to protect ourselves."

In Nanaimo, news reached the union that Western Fuel, idle since May 1st, was starting to work its Number One mine on Monday, August 11th. Word passed quickly through the town. At 7:00 a.m., ten local miners, escorted by mine officials, had to march to work through a crowd of 150 men and women who lined either side of Fry, the street that led to the mine.

The crowd only shouted at them; it was waiting for a fire boss who was just coming off night shift. The union had agreed to allow supervisory staff into the mine, but only as long as scab labour was not used. With ten men back at work, the fire boss was fair game.

The union people followed the fire boss, who pretended to ignore them, as he headed for home. One man shouted, "If you go to work tonight, you old son of a bitch, you will get killed." Along Fry Street to Milton, up four blocks to Victoria Road over one hundred angry people followed the solitary figure. They hurled their frustration at him in insults and curses. When he finally turned and told them to shut up, someone grabbed him. The people surged around him, some joking, some sarcastic. They jerked him backward and forward until a policeman intervened and escorted the fire boss home as the crowd followed shouting, "We're going to take you home, see where you live." They stopped at his gate and watched as he disappeared inside his house. Five minutes later, the crowd had dispersed, but they were back at the pithead when the day shift left the mine.

> We were at Number One shaft when they started to scab Number One. We'd be picketing outside the mine and our union power men would beat them up and would get into trouble.

The scene in front of the office at Number One mine on August 11th at 2:40 p.m. was chaotic. There were legal pickets and additional strikers; there were wives and daughters; and there were policemen – local, special, provincials on horseback. Everyone was shouting – the women loudest of all. The noise was hard on the ears. Police Chief Jacob Neen asked the crowd to leave, but the people demanded that the special constables be sent away. When a blast from the Number One whistle interrupted to announce the end of the day shift, the protesters surged up to the edge of company property, policemen on horseback wedged between them. As the men who had been working got off the cage and headed for a car escorted by police and company officials, the crowd hurled stones at them. Six of the strikebreakers melted into the crowd – the remaining four were members of one family, union men who had been drawing strike pay.

Chief Neen and Harry Freeman, manager of the Jingle Pot mine, who was there to reason with the strikers, packed the four into the chief's car. Inching its way slowly ahead, the car could move no more quickly than the crowd, which kept up with it all the way to the family's house. The four men scurried inside amid a barrage of stones that broke several windows. Then the door reopened to reveal the owner protecting himself with a gun and promising not to return to work.

And the crowd had not forgotten the fire boss. When he came to work at 11:00 p.m. with his police escort, a crowd of two hundred met him at the corner of Milton and Haliburton, four blocks from the mine. The police chief and a constable had gone to his house when he called asking for an escort to work. They had found the fire boss hiding under a bush in his garden. He was right to be afraid. The joking and sarcastic crowd of the morning had become a howling mob, hurling rocks and taunting him and his escorts until they disappeared into company property.

Eight hours later, at 7:00 a.m. August 12th, the crowd was back outside Number One. Their attempts to dissuade a number of fire bosses from going to work having done no good, they were milling around outside company property, their tempers short and their indignation growing with every hour. Among the crowd of two or three hundred that day were a few men who had been drinking and a few more who were fighting mad. And, according to at least one witness, the "union power boys" had positioned themselves throughout the crowd.

At about 2:30 p.m., three miners arrived to work the afternoon shift. The crowd rushed them, swaying them back and forth, kicking their dinner pails, knocking them down. Four policemen rushed in. A rock hit a special. The crowd demanded that he leave and his departure defused the situation for a few minutes until the whistle blew for shift change. The day shift was afraid to face the angry crowd. Someone yelled, "Come up you bloody scabs. Come up and get your medicine." Only when they promised to stay away from work was the day shift allowed to leave.

But still the crowd would not disperse. Finally the manager, Thomas Stockett, accompanied by Conservative Member of Parliament (MP), Frank Henry "Harry" Shepherd, came out to speak to them. The two men proposed a meeting to take place twenty-four hours hence between management and a committee of miners from Number One. From the crowd came a warning, "Fellow workers, you have got these men in a hole. Let's keep them there. They want this twenty-four hours to allow them to get more special police, thugs, and riff-raff from Vancouver, so they can beat us."

But the majority agreed to the meeting. Twenty-four hours later the committee would arrive to meet with Stockett, but among their number would be Frank Farrington, who had come from UMWA headquarters to lead the union effort. For management to have met with Farrington would have been to recognize the union. The meeting would not take place and the violence would continue.

And the afternoon riots of August 12th were not over yet in Nanaimo. One thousand strikers confronted Const. Hannay and six specials at the new Reserve Mine. The crowd smashed Hannay's car, forcing the police to flee the scene. Strikers invaded the office of the editor of the *Herald*, no supporter of the union, and threatened to dynamite his presses. There were hundreds of people in the streets. Then, like water when it finally breaches an earthen dam, three to four hundred rioters spilled south out of Nanaimo along the dirt road that was Haliburton Street and out onto the Black Track. Seven kilometres beyond lay South Wellington.

By late afternoon, they could see the PCC mine even as Superintendent James Roaf saw them approaching from the north. Turning to his left he saw 150 to 200 men congregating around the fence at the south end of the mine property. Inside the fence were the boarding houses and four company houses with their strikebreaker tenants and the carpenter shop that had become the cookhouse. As the superintendent watched in fear, the Nanaimo mob merged with their South Wellington compatriots and moved as a mass across the E & N tracks and into the bullpen, howling slogans and brandishing clubs, Big Louis Nuenthal in the lead.

The men in the cookhouse scrambled out the back door, some fleeing into the woods, others into the mine. Their pursuers knocked the cook to the floor and broke his arm. Gasping with pain he made a break for the back door only to meet twenty-five or thirty men who took him prisoner and marched him to within half a mile of Nanaimo before they turned him loose with instructions not to return.

By now it was dark. Back at the mine, chaos reigned, the night scene illuminated by a large yard light. When Superintendent Roaf tried to confront the mob, they hurled rocks at him. He fled into the dark. Having smashed the cookhouse crockery and windows, the mob advanced on the bunkhouses where it hauled fifteen strikebreakers, four engineers, and a number of Chinese labourers out of their beds and drove them outside. The mob bashed trunks open, tore and scattered clothing, slashed tents, shattered windows. To cries of "sons of bitches," bloody Scotchman," "kill him," "split his skull," and "scabs get out of here," the strikebreakers fled.

They hid in the swamp, they huddled in the bush, one found a cupboard and crawled inside. Those who had hidden in the mine escaped through the airshaft and made their way south. All that long night the fields and woods around South Wellington teemed with the hunters and the hunted. Nineteen mules unaccustomed to freedom added to the confusion as they stampeded through the woods.

> My dad had a little orange-coloured horse named Nellie. She was an old racehorse and was still fast. He and Nellie would patrol the road between South Wellington and Extension on watch for scabs. He'd scout there among the ridges back of Scotchtown and if he spotted anybody that looked suspicious he'd gallop back and tell the boys where they were.

> Around midnight my dad was bringing two visiting Englishmen along the Black Track towards South Wellington. So here's these two Englishmen with suitcases, which the strikers would normally take to be scabs arriving. They were going along the Black Track when these strikers came jumping out with clubs and sticks. "Where the hell are you going?" My dad says, "It's all right" and tells them who he is. I guess these men were marauding around and lookin' for strikebreakers to beat 'em up.

With the mob dispersed in the woods, the company doctor and the superintendent began to search for strikebreakers in hiding. The wretched men came out of the woods, some with blackened eyes, some with torn clothing, but most just badly frightened. The two men opened their homes to them and to the women and children from the company houses who had fled into the bush as well. Other homes in the neighbourhood became places of refuge too.

> A friend of mine was farming then and he put up an injured scab overnight. A mob came up to his house and demanded that any men inside give up. They threatened to burn his house and barn but he stood up to them.

> A bunch of strikers found this scab, who had been beaten unconscious. They brought him around and helped him back to camp.

The police were outnumbered and powerless to control the mob that night in South Wellington, even when reinforcements arrived. They were able to identify some of the participants, however; they recognized Louis Nuenthal, the men from Nanaimo, and a number of outsiders who appeared to have organized the riot in advance.

Thirteen kilometres farther south on the same night, five Ladysmith policemen watched the streets nervously as two or three hundred strikers milled around outside the Temperance Hotel. It was 11:00 p.m. and the air was still very warm. The union had issued a warning in response to the Hatfield stabbing three nights before: "If the police do not protect us, we will protect ourselves." The five policemen knew they wouldn't be able to do much against three hundred strikers even with the reinforcements that had arrived earlier in the day.

> During the day, a number of special policemen had been brought into Ladysmith, some mounted. Feeling was quite stirred as one of the mounted policemen was accused of being under the influence of liquor and of riding his horse on the sidewalk, injuring a woman.

Local president Sam Guthrie had done everything he could to keep things calm in Ladysmith. The loading docks were a particularly troublesome place. One night Sam had arrived just in time to stop some strikers from pushing a small hut full of strikebreakers into the ocean. The loaded railway cars on the same dock were another favourite target. A group of angry men could tip their contents into the harbour easily.

On August 12th, earlier in the day, a boat had slipped quietly into Ladysmith harbour. On board were Patrick Fagan, the Canadian Collieries paymaster, and nine drunken strikebreakers. Fagan warned them as they landed that they would have to go through picket lines to get into the hotel and should refrain from speaking to anyone lest their drunken state make them belligerent. As they approached the hotel, however, it was their very refusal to speak that angered the strikers.

Fagan had noticed that Charles Axelson, a Swedish-born striker, had followed them up from the wharf. The next time anyone noticed Axelson was when he came out of one of the local bars having spent enough of his weekly strike pay to make him drunk. He staggered up the street singing, "I belong to Glasgow, dear old Glasgow town."

> The pedestrians laughed at him and some of them patted him on the back encouraging him to keep singing. But when he suddenly started singing the strikers' theme song, "Hurray, hurray, we'll drive the scabs away," it was bitterly resented by the non-union men.

The strikebreakers complained to the police, who took Axelson away to jail still singing. But the little Swede had a formidable wife who was the president of the women's auxiliary to the union. Mrs. Axelson, "a veritable Amazon in build, vigour, and strength," according to Lempi Guthrie, Sam's wife, wore old-fashioned clothes and a bashed-in man's hat. She was chairing a meeting when word came of her husband's arrest.

> She picked up an axe that was at the back of the hall for chopping wood and went to the back door of the jail. She demanded that the policeman release her husband. When he told her he wouldn't do that she showed him the axe. I guess he figured she meant business so he rushed into the cell and let Mr. Axelson go. When they came out there was a fairly large crowd so they marched down to the Temperance Hotel with the Axelsons in the lead.

The situation in front of the Temperance was much less jovial now. The crowd threw rocks and pushed and yelled and sang, "Hurray, hurray, we'll drive the scabs away."

In the course of the evening there was a lot of milling about. I spent most of the evening and night on the steps of the Canadian Bank of Commerce, cross corner from the Temperance. Later in the evening a blast went off at the rear of the hotel. It was said no one was injured, but there was a lot of glass broken.

Shortly after twelve o'clock there was a loud explosion in the rear of the hotel. Someone in the crowd shouted, "Hurrah. The balloon has gone up."

The crowd moved away from the hotel to seek other targets for their anger. Shortly after midnight they found what they were looking for at the home of Alexander McKinnon.

There was a man in Ladysmith. McKinnon was his name. And he went to the union meeting and he says, "Boys, it's no use stayin' out any longer. I've got to go back to work." So he went back to work.

Mr. McKinnon had no strong feelings regarding the strike but he made the mistake of tantalizing the strikers. When coming home from work at the mines he would open his lunch bucket and offer the crusts of his lunch to the picketing strikers, bread that had been bought with money he had received from the union before he went to scab. Naturally this did not sit well with the strikers although they did not take him seriously enough to do anything about it.

There were two fellows who weren't even strikers but they had been talking and trying to get someone to take more drastic action to end the strike. Stackhouse was a barber and Tangle Jackson was a bartender. They collected some dynamite, which they carried to the top of McKinnon's house. They lit the fuse and threw the dynamite into a bedroom that happened to be occupied by the children.

McKinnon heard a noise and rushed into the bedroom. He picked up the dynamite and threw it at the window but it hit the curtain and bounced back. He picked it up again but the dynamite exploded and blew his hand off and one eye and mashed him up against the wall.

This incident reflected badly on the strikers in the outside world. My husband was Sam Guthrie, the president of the Ladysmith local UMWA. Sam felt very badly about the matter and later he and McKinnon became devoted friends. [McKinnon] regretted what he did in later years and counselled young men never to scab.

With regard to Stackhouse and Jackson, Stackhouse turned King's evidence and escaped without punishment. Jackson received seven years in the penitentiary.

Feeling was running very strong against Stackhouse and there was a good chance he would be killed by some of the strikers if he returned to Ladysmith. Consequently some of the strikers managed to get a launch into

> the harbour, gave Stackhouse $50 and told him to leave for the mainland and not come back.

> There was a strong rumour that Tangle Jackson didn't do it. McKinnon later tried to get him released from prison.

The shock of the second explosion and the sight of a badly wounded man did not shake the determination of the crowd. They divided into two columns, one led by William Stackhouse and the other led by Joshua John "Joe" Taylor, an official in both the British Columbia Federation of Labour and the UMWA. It was almost morning as the two columns set out to march through Ladysmith and visit the homes of men who had chosen to return to work, men who lived with their families in their own houses scattered throughout the town.

> By now it was August 13 in the wee small hours. They went to the houses of a number of strikebreakers. They were trying to dissuade them from going to work. A number of them had their houses damaged by rocks thrown through the windows.

Ladysmith, named to commemorate a town under siege in South Africa, became a siege hostage itself as some of its own citizens marched its streets and others defended their homes by brandishing guns.

> I heard the two explosions but I did not go out. Next morning a crowd of strikers came to our house throwing stones. They smashed fifteen windows in front and eight in the rear of the house. My wife hid under the bed for safety against the rocks coming in. Then they moved to someone else's house and treated it the same way.

> I can remember as a little kid, before the riot started, this whole mob of men being at our door and my father talking to them. And they said, "Davie will you come with us, it gives us respectability." And Dad said, "I'm sorry. You're wrong. You mustn't go on strike. It's the wrong thing." And they said, "Well, there's going to be trouble tomorrow and your home is one of them." And they did wreck our home; they wrecked several others along the street including McKinnon's. You could see the mob coming.

The early morning found the marchers in front of the Temperance Hotel again, their anger still fuelling their thirst for revenge.

> Six o'clock the next morning two very worried looking strikers knocked at Sam's door. Apparently the strikers had gathered again and were still on the warpath. Sam thought he would be able to influence them to go home.

After warning his wife, Lempi, to stay inside, Sam left for the Temperance Hotel. Shortly afterward, a procession of men and women passed the house singing. Lempi decided that Sam should know, and she hurried after him, catching up with him at the train station where the strikebreakers were boarding the train to go to work.

> A burly Scotch miner had a strikebreaker down, head hanging over the platform. With his hands on the man's neck he shouted, "Ye whore. Are ye ready to dee?" I rushed up and tried to loosen the grip on the man's throat. "Off wi' ye, ye blasted imp." he yelled. Then he looked at me and said in a milder tone, "'Tis no place for a bonnie young lass like ye; gae hame wi' ye." Just a little further along, two tiny Scotch women had grabbed the dinner pails from the nervous hands of two husky young fellows and were smearing their faces with the jam sandwiches. "Now, ye blasted ninnies, go hame and see if your ain mithers will ken ye," they shouted at them.

Lempi said that Sam had spent the night at home and police Const. A.C. Pallant would testify in court that Guthrie and Taylor spent the morning helping the police to restore peace. But strikebreaker Owen Dabb swore that he had seen them from the window of the Temperance Hotel half an hour before the first explosion occurred.

All that day, Wednesday, August 13, 1913, bands of strikers continued to march. One disorderly group descended on Chinatown, damaging and looting stores and boarding houses. Other groups were more disciplined. Marching two-by-two in columns, they revisited the houses of strikebreakers. This time they did not hurl stones and shout angry words. This time they argued and cajoled, quietly but forcefully.

> Later my dad received a visit from about forty men. He was ordered to leave town but if he would sign a guarantee, the union would protect him.

> All those visits did was scare people even more. The people whose windows were broken went to Victoria by train after being threatened.

The Ladysmith riot did not persuade strikebreakers to come back into the union, but it did prevent them from working for a while. Fear, not reason, had prevailed. Just as in Nanaimo and South Wellington, the Ladysmith riot was organized by the union beforehand as a desperate measure to force a settlement.

> I guess this had been planned ahead of time by all the union officials and all those that were in the union and they were going to have this demonstration downtown. . . I know it doesn't sound kind and I don't mean it unkindly when I say it but those who did take part in it were either very young or they were not of an educational or intellectual background to quite realize the whole picture.

> I talked to one of the Ladysmith rioters. He was very bitter because he says he well remembered the day before they were being stirred up by all these leaders and what they were going to do and the riot and so on. And he said, "Guess where all the leaders were? Out in the bay, quite visible in boats, fishing. And we were the suckers." And I said, "Well, Johnnie, how old were you?" And he said, "Most of us were fifteen or sixteen years old." And this is a fact that should be brought out, that these were just kids.

There were teenagers in the thick of the riot, but most of the participants were adults. It would appear, however, that local union officials were not party to the plans of the international organizers. Sam Guthrie, in particular, though young, was a man of character and principle.

> Sam was very opposed to violence. In many cases he was instrumental in avoiding violence on the part of the more hot-tempered strikers. The fact that many incidents were provoked by the riff-raff brought in by the company made no difference to Sam. He felt strongly that reacting by violence was wrong in itself and also would do no good to the cause of the strike.

> Sam Guthrie was one of the leading strikers. My father was not a striker. He had been through the strikes in Old Wellington and he wouldn't have touched it with the longest stick he could hold. So on that particular very sore point they were miles apart. But when Sam retired, he spent all his time sitting in the back of the liquor store chatting with my father. And he told me then he thought it was a big mistake. He didn't recognize the American influence on this local situation.

Sam Guthrie went to prison for his pains. He and Joe Taylor received two-year sentences – the longest of anyone. When Sam got out of prison in the general amnesty of August 1914, he found public sympathy to be with him and he turned it into a political career. When Parker Williams became a member of the WCB, Sam succeeded him as MLA for Newcastle.

> Sam Guthrie used to come over to our house quite often. During the strike there they had these special meetings and everything. I remember him sitting in our front verandah time and time again and he and Dad'd start these political arguments. Each time those guys got together they'd start arguing politics, politics and more politics.

> Sam Guthrie was a good man. He was a man who was there for the working man. I remember when Sam was in the House down there. He was always asking for steady hours. An eight-hour day, a forty-hour week. He wanted a thirty-hour week so more people could be put to work. Off the unemployment rolls. And his idea was good.

While the Ladysmith strikers under the steadying influence of Sam Guthrie chose reason as their second line of offence, South Wellington men retained their tough stance of the night before and spent the morning of August 13th rounding up the scattered strikebreakers in preparation for shipping them out of town. The superintendent of the PCC mine described his confrontation with the union men.

> At 11:00 a.m. a number of men came up to the mine and I spoke to their leader, Louis Nuenthal, and he said they wanted these men to take them away. There were sufficient people to force their demands if I refused. He said they wanted Dempsey especially and I went over to the house and told

> Dempsey and he, together with some of the special policemen, came over and gave themselves up. Mr. Tompkins, president of the company, arrived by train from Nanaimo about that time and he had the train stop at the mine office instead of the station. That brought a large crowd of about two hundred men and a few women on the run armed with clubs. I guess they thought he was unloading more strikebreakers. Tompkins ordered them off company property.

Louis Nuenthal and his cohorts roughly herded nineteen strikebreakers into the station waiting room, knocking some of them down repeatedly. A striker named Prendergast put his head in the window and shouted, "If you ever come back to the Island, we will kill every damn one of you." Dempsey, a bullpen resident for two months, was bleeding freely from a head wound. In one of the only compassionate gestures witnessed that morning, one of the strikers went for a bucket of water and bathed Dempsey's head with his handkerchief.

Disheveled and battered, the nineteen men waited one long hour in the hot little waiting room until the E & N train from the south arrived. Nuenthal took up a position at the door leading to one of the coaches, and his followers herded the strikebreakers, carrying what belongings they had been able to gather, on board.

The train arrived in Nanaimo a few minutes later and disgorged a subdued group of men, the violence and sleeplessness of the previous night having drained them of everything but the wish for it all to end. But there was more in store for them. As they marched down Fitzwilliam Street from the E & N station to the CPR wharf where the S.S. *Princess Patricia* was due to dock around noon, two of them tried to break away. A crowd of about five hundred miners surrounded them, foiling the attempt by quietly pushing them back into line.

But the scene on the wharf was not as restrained. When the steamship docked, forty or fifty strikers met twenty-five special policemen coming down the gangway and pushed them back up onto the ship with the warning that anyone who attempted to come ashore again that he would find himself in the water.

Then the crowd turned to a provincial policeman who had been observing the incident from in front of a shed. Even though he had hidden his badge and his gun underneath his jacket, his civilian's suit and white straw hat stood out among the roughly dressed crowd. Someone grabbed him and pulled him into the shed shouting, "Come on fellows, we've got a special. Kill the son of a bitch. Kill him."

The policeman shook himself free and pulled his gun. All his attackers stood back except one man who shoved his hands up and came towards the law officer saying defiantly, "Shoot if you dare. I'm not afraid of you. I've been to the Boer War." Just then another constable appeared and the cornered man lowered his gun. In a flash, the crowd attacked; the strikers seemed crazy as they beat him. MLA John Place, who happened to be there, tried to reason with them and took a revolver from one striker. Place paid for his efforts with a jail term.

The South Wellington men pushed their prisoners through the crowd and up the gangway of the *Patricia*. The captain gave the order to sail. The ship pulled away, the specials and the strikebreakers perched on the bow staring back at the crowd whose numbers had swelled to a thousand persons.

Now began the rioters' exodus to Extension. Later, groups would leave in response to news of a battle there, to word brought to the *Herald* office that six strikers had been killed, but the men who left in the early afternoon, especially the South Wellington group under Nuenthal, had to have been acting according to a pre-arranged plan, because it was too early for anyone in Nanaimo to know what was going on in the coal camp behind the ridge at Extension mine.

The trouble had begun to brew on August 12th, the day before, when residents and company officials saw union organizer Joe Angelo pointing out strikebreaker's houses to two other men. Knowing what had happened in Nanaimo on August 11th, company officials moved all the strikebreakers inside the bullpen for protection. Word had come from the provincial police that reinforcements were arriving from Vancouver on the *Patricia* the next day. Strikebreakers and management alike hoped they would arrive in time.

Fear stalked the Extension bullpen that night. Whether to intimidate the strikers or to give the men in the bullpen courage, a searchlight shone on the strikers' homes. And false courage poured from liquor bottles.

> They had a bullpen for the imported scabs where the men were guarded by special police. That was why they were so cocky. They had a drinking party and said they were going to chase the union men out of Extension.

> Well you know drunken men. They'd be howlin' and shoutin' and Extension's a small place. The mine is right here and houses all around. You could hear them. They were having a party and I guess the booze was talkin'. They were going to clean the union men out of Extension. So my brother and some other young fellows laid out all night with rifles waitin' for them.

Very few people got any sleep that night. If the strikebreakers in the bullpens were nervous, so were the union people. Rumours spread quickly. Most people believed the one that said the company had given its men guns and ammunition.

On August 13th, after a night of watching, everyone in the camp was ready for anything. At 7:30 a.m., they could see Joe Angelo with a group of men by the post office: he seemed to be pointing out houses again. The camp may have been cut off – there were reports that the telephone lines were down – but somehow up to six hundred Nanaimo and South Wellington men knew they were needed in Extension that morning.

> We were staying in the Cinnabar Valley at that time and they marched up. They came through the valley on a trail past our place to go to Extension.

> At that time we were going from Nanaimo to South Wellington to the bullpens there and we were away out by Chase River when a boy came up on horseback from Extension. He was heading for the union hall in Nanaimo to tell them the scabs had started on the union men. So we decided to go up to Extension instead.

> Well, if you've done anything on the strike you'll know that they sent the goon squads in with guns to Extension.

> So I got a bunch of guys to come up from Nanaimo. Horses and wagons, they all came out. A lot of people with their clubs and everything.

As morning became afternoon, the word spread through Extension that a big crowd was on its way. The wives and children of the strikebreakers retreated into the bush. Some say they were most afraid of the union women.

> I can remember seeing the strikers gathering on the hill where the road comes into the town. The women were screaming so I sent them into the bush and went to the company office.

> The people from Nanaimo seemed to meet the union men from Extension before they rushed the pithead.

> The crowd was coming down the hill. The hill was black with them.

When hundreds of people are caught in an angry confrontation, with bullets flying, and women screaming and men running for shelter, regrouping, running to another position, the confusion generated produces as many versions of the facts as there are people involved.

> The union gathered that day on top of the hill in Extension and sent word over to the mine where the scabs and special police were all congregated to pick a committee from each one and they would try and settle something. Anyhow, this special police he come over from the mine and brought word that they wouldn't talk. Now I didn't see it, but if I would have hurried up I could have seen it, 'cause I was running right up the hill. I was at the post office at the time, but as I was running up the hill, the bullets from the mine – they were shooting at the strikers because someone had hit the special police on the head . . . Now whether they seen that from the mine, 'cause there was quite a distance . . . Now the miners had nothing, only baseball bats, that's all they had, but the firing started from the mine, in fact one fellow got shot.

They all had guns and they all come up to Extension. Then the war started. I was right there. I could see the tipple for Christ's sake. Hear the bullets hitting the tipple.

Fifty or sixty men were shooting at us up on the ridge. We shot back. The bullets came past; I could feel the vibration going past; you could hear the bullets.

Buckshot was coming right out of the pithead.

My grandparents lived in Extension and the strikers were lined up on what we called the Bluff with guns trying to kill my grandfather. His sons were firing back with rifles while my grandmother and aunties and my oldest brother, who was a little shaver, escaped into the woods.

The union men tried to get near the mine but bullets were coming out. But the scabs never tried to kill. They fired over their heads just to scare 'em back, If the scabs had of fired into the men they would have massacred them 'cause they was all just standin' there in a big bunch and they were gonna wreck the place. And things went against 'em.

There was two or three hundred people came on the tipple with clubs and sticks and then there were a few shots fired and then the bunch ran away. I stood on the tipple for a while and then went on to the ridge until 6:30. Shots were fired from both the tipple and the ridge. I could see the strikers were getting around us all ways so we left the ridge at half past six and then we went on the narrow gauge and stayed there until 8:00.

Bullets or baseball bats. Each side accusing the other of being the one with the guns. Six men dead in Extension. Seven men dead in Extension. Nine. Even after the shooting stopped, no one knew what had happened for sure.

It was war. In fact, a fella by the name of Alec Baxter, he was a contractor, not a miner. And the scabs was in the tunnel out of the way, you know, and they shot Alec Baxter. I think he was the only man got killed though.

Alex Baxter was up there, a spectator, and a bullet come out of the tunnel and hit him. He was the only man that got hit with a bullet. Hit him in the side.

To my knowledge I don't think anyone was actually shot, on one side or the other side. Killed anyway. The shooting was mainly from a distance.

And there was one fellow from Nanaimo come up there and he walked across the front of the mine and suddenly a shot come out and it hit him in the stomach. His name was Baxter. He didn't die but that's the only casualty as far as maiming is concerned. Well, a lot of them got beat up.

So, anyhow, when a bunch of lads from Nanaimo started shooting, the scabs got back in the tunnel of the mine. So Baxter, he was a contractor in house moving. After the scabs quit shooting, he went down in there. When he

> passed the mouth of the tunnel he got shot but it didn't kill him. He died after from the effects I suppose. After that the scabs walked out of the shaft.

The shooting began around 3:00 p.m. and continued until at least 6:30 p.m., possibly later. While the strikebreakers in the tunnel and the union men on the ridge were exchanging shots, other groups of strikers were parading through Extension. At least one of these groups consisted of teenaged boys who had worked in the mines for only a few weeks or months when the strike started the year before. Their young heads were full of union bravado and they were ready to teach the scabs a lesson. Sporting white markers on their hats and waving a white flag, they went from house to house. At each one, the leaders pounded on the door and finding no one home – all the occupants having fled to the bush – they waved the white flag and, bent on pillage, rushed the house.

In the aftermath of the riot it became very difficult to determine the identity of any of the looters. Witnesses contradicted one another. Strikebreakers swore they saw boys they had known for years leading the crowd. A striker's sister said he was at home all the time and had even helped the people whose houses were attacked. Sixteen-year-old William Bowater, Jr., was singled out in particular but was later acquitted. The version favoured by the union blamed it on outsiders.

> The fellows came up from Nanaimo when they thought the others were in trouble. There's always a bunch follows. And they did a lot of damage. There were houses set on fire, smashed up and a bunch came in looting too. The miners got blamed for every thing. Most of the miners were good honest people.

Extension's Chinatown was a particular target. Here, in boarding houses that held ten to fifteen men each and in individual homes where four or five lived co-operatively, dwelt 150 Chinese. Years of anti-Asian feeling drove the mob of one hundred or more armed with picks, shovels, sticks, and guns that went from house to house calling on all inside to leave Extension or be shot.

> I saw a large crowd of men armed with guns coming towards the store. They were shouting. I locked the front door and sat down in the parlour behind the store. I heard shots fired; I heard the front of the store being smashed in. Then some men came into the parlour and pointed guns at me and said, "Suppose you not go, I shoot." I got scared and ran out by the back door and hid in the bush. At nighttime I went to Nanaimo. I travelled along the wagon road and whenever I heard a noise I hid in the bush, and then when it was quiet I would get on the road again. About two miles out of Extension two men coming on bicycles caught hold of me and searched my pockets and robbed me of what money I had on me, $2.50.

No one knows how the fires started. On a hot August day in that tinder-dry camp where the wells routinely dried up each summer, it would not have taken much to start a fire. The wind blowing that day only added to the likelihood the fire would take hold. The fire was very selective, however. Eleven houses – cabins or shacks – all belonging to strikebreakers, caught on fire, including all the houses in the bullpen. Added to these were several mine buildings and the manager's home.

> The first we saw was the end of the tipple, the houses burning down, the motor shed, and then some more houses and then there was those close inside the company's fence and then we saw another house burning.

> I heard shooting between three and four o'clock. When I arrived at the top of the hill – it is very high, you can see all over Extension – there was a lot of people hollering and smashing windows and doors of the houses and there were a number of places on fire, dwelling houses and the car shed. It's a long shed. The trestlework and tipple and weigh house is all connected in one. This was all on fire.

> I saw one man playin' a piano with an axe.

Thomas Isherwood, a striker and store owner from Extension who had been working at another job and living alone in a tent away from home, was walking back to Extension along the wagon road when someone told him there were riots in Extension and seven strikers were dead. His wife and two of their children, Ethel and Tom, were with him, having come to bring him food. Their daughter, Alice, was at home tending the store and three more of her siblings. Two days later, in words spelled to reflect his origins, Tom would write to his adult daughter, Emily, in Victoria.

> They told us their was no more Extension has it was all burned down. But has we came along we found that not to be true. For we met a man has told us the Union men was looking after our house. You may bet we was in an awfull way. Tom kept crying and Mother and Ethel was near crazy and I was not so far off I can tell you. So when we got to the top of the hill we could see that our house was all right but their was fires all around. So we pushed our way along through the people of which their was over a thousand lotts of them with guns until we got home and we found everything was all right only Alice and the babys were all crying and Alice has not got over it yet. Mother, Alice, and Ethel never went to bed all that night. But I got Tom and Annie and Babys to come and sleep with me has I was tired out with working that day and then having that long walk home.

The story about the boy on horseback spreading the alarm seems to have been true, although the news he shouted was false. But the false news had spread rapidly and the details had changed. The August 13th edition of the *Free Press* shrieked: "STRIKE SITUATION ASSUMES SERIOUS ASPECT; NINE MEN SHOT IN EXTENSION TODAY."

Nanaimo was ripe for rumours. Having itself been the scene of a riot at the CPR wharf at noon that day and having watched the anger of the strikers at Number One mine at 3:00 p.m., when Manager Stockett objected to Farrington's presence on the strikers' committee, Nanaimo now saw hundreds more of its citizens preparing to march on Extension.

> I was near Quennel's butcher shop on the main street and I saw a group of strikers with guns. While I was talking to a man there, a man came on a motorcycle and said, "Six men have been killed at Extension." The strikers marched off with guns on their shoulders along the street towards the fire-hall. I returned to Chinatown and six of us set up a guard and walked up and down all night watching.

The *Free Press* reported seeing a large crowd of men passing firearms out of the door of Hilburt's store. Another witness observed men going to various stores looking for guns and ammunition but coming out empty-handed as the authorities had removed all such merchandise and ordered all the saloons closed too. But some of the crowd that left Nanaimo after 5:00 p.m. were armed either with stolen weapons or their own hunting guns. They left on foot and on horseback, rigs loaded with men overtaking the marchers, an ambulance making a speedy exit south. Some of the rigs carried UMWA officials; some carried provincial police; all were heading for Extension.

> My dad was a fire boss and we lived out in Chase River. It was a summer day and my father had been working and when he got home we were bucking wood and these two men came along and they said, "You'd better get away from here with the boy because they're rioting and they're going to come up here too." And they were howling and hooting. They'd raided stores downtown and got all their guns and ammunition and then they went up to Extension mine. Dad said, "You go back and tell them that I'm not gonna run. I'll be behind a stump and if I go down, there's gonna be someone else go down with me. You go back and tell them." He just kept sawing wood. My mother was worried the whole time. She couldn't do anything else but worry.

Afterward, someone would estimate that there were fifteen hundred people in Extension by the time night fell. The normal population was about three hundred. With darkness, the shooting stopped as people scattered, some to hide in the bush, some to patrol the road, some to fight fires that burned in the buildings around the mine. They extinguished the pithead fire once, but it would come to life again by the morning. The strikers had dismantled the company mechanic's automobile and set it afire. Coal cars burned and so did eleven houses.

> Them shingles was comin' down there pret' near a whole shingle on fire, right through the wasteland. When they get up in the air they fly for miles. We just got the hay cut, see? The whole family was out all night watchin' the

hay didn't catch on fire. They burned a dozen houses I guess. The burnin' houses was all scabs'. They didn't burn any union houses. They cleaned out lots o' stuff up there before they burned 'em. They put it in their houses; they know what they're doin'. Somebody saw me in the bush there. He had a full bed. A Scotchman. He says, "How's your family fixed for beds and have a look at this one." It was one o' them fancy beds, big headboard, everything else.

A great deal of plunder was brought into South Wellington from Extension. In Extension it got pretty rough. They burned the superintendent's house down and one of our neighbours the next day was wearing his Panama hat. My dad says, "Look at him going down there, that's Cunningham's hat."

Cunningham, the manager, had a lovely dog, a pointer. And he'd come around the house and my mother fed it. So my father and these two other fellas were taking a walk along the back road after the riots with the dog and this fellow was hiding, a scab, and he saw the dog. And he thought it was the boss and he come out. And they grabbed him and brought him up to our place and put him in a room and my mother fed him. Nobody touched him or put a finger on him. But yet he got up in court and said my father threatened to shoot him.

The terrorized wives and children of the strikebreakers spent the days and nights that followed huddled in the bush.

I was fourteen years old and my father had been working for the company. I was coming home from Nanaimo in the grocery wagon. A large group of men with clubs stopped the wagon and one of them says, "This is a scab kid." They turned the wagon around and told me to go back to Nanaimo but as soon as I could I got off the wagon and went into the bush.

I owned a store in Extension and I wanted to see what damage had been done so I hired a buggy in Nanaimo for $3 and drove to Extension. Not far from there I saw an Italian lady I recognized and three more white women. One was carrying a baby. When they heard the buggy coming they ran into the woods. I took off my hat and spoke to them. They were crying and didn't speak.

During the strike there was a lot of trouble. When they burned the houses in Extension, a lot of owners were Italian and some came to my dad's hotel in Ladysmith, and when the union men found them, they just beat them up. Blood was flying. It was terrible. It was really cruel and there was wives with them too.

On August 15th, Thomas Isherwood told his daughter in Victoria that there were no scabs left in Extension. Their houses were burned, their possessions scattered. "There is all kinds of stuff lying around." Horses and pigs roamed the town with no one to look after them. He had decided to stay home for a week until things were quieter. He wrote that the strikebreakers had been arrested and were in Nanaimo. He wrote that the reason there were soldiers in the town was because they had come "to get the scabs out."

The provincial police took eighty adults and forty-two children out of Extension and fed and housed them in Victoria. Heavy traffic plugged the road as soldiers escorted other women and children in wagons to Nanaimo. Soldiers on horseback passed them, heading for the ruined coal camp.

Soldiers were moving into all the coal camps on the Island. The rioting had spread back to Cumberland, where the first altercations had occurred. On August 14th, white strikers threatening Chinese and Japanese homes had "cowed" the police, according to reports, and dissuaded them from taking action. Later that night, a gang of strikers tried to set fire to the wharf at Union Bay. Although the police had charged the men and put out the fire, things were clearly getting out of hand. When two men attempted to blow up a railway trestle at Union Bay, the police asked for help. Help came in the form of the militia.

CHAPTER VIII

Bowser's Seventy-Twa

> Then they brought in the militia to Cumberland to break the strike. Nanaimo too. The guys were paradin' and the militia beat the shit out of 'em. Yeh, they were tough old days then. – Wink English

> And the people in Nanaimo said, "Oh by God, you'll sure hit dynamite if you go up that road to Extension. There's all dynamite laid out for you, you'll get shot up." You know four or five fainted on the road going up to Extension. Sure they did. They scared the devil out of them here in Nanaimo. – Ellen Greenwell

> They'd been workin' on the road and it was all dug up and the soldiers thought it was mined. – Anonymous

> Well all that trouble led to this. They brought a boatload of Bowser and the Seventy-Second militia. A Bowser is a member of Parliament in Ottawa. They were all in kilts. – Harry Ellis

The miners of Vancouver Island remember William John (Bill) Bowser, attorney general and acting premier of British Columbia, as the man who called in the militia. In response to a request by Nanaimo's Mayor Shaw and two magistrates, and in opposition to the wishes of the strikers, who promised to preserve the peace if he withdrew the specials, Bowser ordered in militia units of several army regiments – the so-called "Aid to the Civil Power." Their job was to support local police in their efforts to restore peace.

Except for a sprinkling of Boer War veterans and former British army regulars, the militiamen were enthusiastic amateurs: Sunday soldiers with no experience, little military training, and no training whatsoever in riot control. Most of the regiments they belonged to were newly formed. But Bowser's action reassured the general public, alarmed as they were by press reports that said the riots were the work of foreign agitators. There were some people, however, who said that all Bowser was doing was granting yet another favour to the owners of Canadian Collieries, railroad magnates Sir William Mackenzie and Sir Donald Mann.

Because Bowser's main concern was that the operation be conducted quickly and in utmost secrecy, the commanding officer, Lt.-Col. A.J. Hall of the 88th Regiment, Victoria Fusiliers, took elaborate measures to comply. He arranged for a special E & N train to act as a decoy and allowed the news of the train to leak to the press; then he rallied his troops in Victoria, marched them on to a steamship, and sailed for Departure Bay in the middle of the night. Early on the morning of August 14th, one thousand fusiliers in full uniform disembarked at the Brechin mine and marched into Nanaimo.

Later that day, in Vancouver, the 72nd Regiment, Seaforth Highlanders of Canada and the 6th Regiment, Duke of Connaught's Own Rifles (DCOR) marched through the downtown streets to the docks. The 650 soldiers, especially the kilted Highlanders, looked well prepared and impressive at first glance, but a careful observer would have noticed the random straps and pieces of string that held equipment to belts and the boots that had come not from the quartermaster's stores but from a closet at home. On either side of the streets, the taunts and catcalls of spectators made it obvious that the enterprise was no longer secret. Adding to the tumult of the farewell was a short-lived sympathy strike by the crew of the S.S. *Princess Patricia* that greeted the soldiers on the dock.

That the operation had been mounted in haste was evident in the lack of sleeping space on board the *Patricia* as it sailed through the night for Union Bay and in the absence of tents, blankets, cooking utensils, and food at the makeshift camp in Cumberland when the soldiers arrived in the morning on August 15th. Having commandeered the football grounds and a new school nearby, the militia posted sentries, dug latrines, set up signal lamps, and went to sleep in the school with only their great coats for bedding. By this time one company of soldiers had returned to Union Bay to patrol the coal-loading facilities.

The next day, Saturday, the soldiers supervised as five hundred Cumberland strikebreakers received their pay watched closely by groups of strikers. With soldiers on every corner and all the saloons closed within a fifty-mile radius of Cumberland, the strikers were a docile lot. Then, as suddenly as the soldiers had come, all but a detachment of Highlanders disappeared. They rose to reveille at 2:15 a.m. on August 17th, drank a cup of coffee, climbed aboard the train back to Union Bay, and sailed on the S.S. *Charmer* for Nanaimo to join other units of the militia.

The military was much in evidence in Nanaimo. At first the soldiers had slept in freight cars and eaten at local hotels and restaurants that were forced to provide food. By August 17th, however, they had settled into sixteen-man tents pitched closely together in two separate camps – one at the E & N station on the hill above the town, the other down by the harbour in Dallas Square opposite the post office.

At first Col. Hall had consulted with Mayor Shaw, Chief of Police Neen, MP Harry Shepherd, and the attorney general's representative,

T.B. Shoebotham, but the consultations ceased when their conversations became common knowledge. The colonel replaced telephone and telegraph operators with his own men when he learned that the strikers were privy to his messages and troop orders. And he assigned two Highlanders to transmit in Gaelic when wiretaps intercepted his messages. He posted fifty soldiers and three officers to stand guard whenever the *Patricia* arrived in the harbour and sent small contingents of militia to guard Ladysmith, Extension, and South Wellington.

Extension had been high on the army's list of trouble spots. Nineteen soldiers had been dispatched there on August 15th under Capt. Eric W. Hamber amid rumours that fifteen hundred armed strikers lay in wait for them on the road to Extension and that strikebreakers were holed up inside the mine. Instead the soldiers found refugees – the terror-stricken wives and children of strikebreakers – hiding in the bush. The soldiers rescued the strikebreakers and their families, sent them to Victoria to the care of the provincial police and returned to Nanaimo.

Soon some of the strikebreakers drifted up to Cumberland to work in the mines under the protection of their rescuers. The riots had not stopped Canadian Collieries from mining coal. On the Monday following the arrival of the militia, the Cumberland mines reported an exceptionally good output, unusual for the first working day after payday.

On August 15th, the previous Friday, miners and businessmen in Ladysmith had asked Col. Hall not to bring in the militia – they would keep the peace themselves. Next morning 130 troops from the 5th British Columbia Regiment Canadian Garrison Artillery had arrived by special train from Victoria under the command of Lt.-Col. Arthur W. Currie. As miners watched from the opposite side of the street, the soldiers pitched camp in front of the Abbotsford Hotel.

Soldiers continued to arrive in Nanaimo – the Royal Canadian Artillery, the 19th Canadian Army Service Corps, the 18th Field Ambulance, detachments of the 6th Field Company Canadian Engineers – until there were one thousand of them. Nanaimo was under siege from within. Tents, arms, ammunition, a machine gun mounted on a railway flatcar, soldiers bearing rifles with bayonets fixed, officers wearing swords – military law was in effect.

> Remember all them soldiers they had around here? You couldn't stop on the street to talk to anybody. Move on. Move on. They had them special police too y'know. Tell ya to move on. Move on.

> The militia closed the bars when there got to be too much rioting and things got unruly. They brought in a law to stop the liquor.

> I was a kid goin' to school. I can remember they got these little huts where the man stands in and we hadda go past them. They had a sentry there

> watchin' so the men couldn't go near the mines. But us kids could go by. Here was a man in a Canadian army uniform standin' guard there. I'll always remember that.

Col. Hall forbade a waterfront meeting of the union, but the regular UMWA concert went ahead as planned. The coal towns waited to see who, if anyone, would be arrested. Then the Jingle Pot mine settled with the union.

Unable to continue in business and influenced by the liberal attitude of its manager, Harry Freeman, the Vancouver-Nanaimo Coal Mining Company, owner of the Jingle Pot, had agreed to almost all union demands. On the evening of August 18th, twelve hundred miners gathered in the Athletic Hall on Chapel Street not far from the army camp in Dallas Square. They had come to discuss the Jingle Pot settlement.

While the miners listened to Harry Freeman and Frank Farrington speak, the militia quietly surrounded the building. The soldiers positioned an automobile facing each exit so its headlights would blind anyone trying to come out. When the meeting adjourned at 11:30 p.m., the colonel walked into the hall. He was followed by his troops, who divided the miners into groups and instructed them to leave the hall and walk between a double line of troops with bayonets fixed and rifles loaded, down the street to the courthouse, where special constables waited to search them for weapons. The police were looking for forty men in particular, and they had arrest warrants for them all.

> Col. Hall said, "Six at a time single file. Any man running will be instantly shot or bayoneted." So I went out in the second group and there was soldiers on both sides with fixed bayonets and a special policeman grabbed me by the arm and we were all taken into the courthouse to be searched for weapons. They took our names and arrested forty-three men. The rest of us were turned loose out in the yard. And from the courthouse there was a row of soldiers with fixed bayonets. We were searched outside again. So it was 3:00 in the morning before we were all free to go and we had to walk all the way home to Northfield.

The colonel's plan had been to assist the police in the arrests and keep the rest of the strikers confined until the militia was in position to assist with arrests in all the other coal towns. He had forbidden anything but official communications after 3:00 a.m., when the strikers at the courthouse who had not been arrested were released to join their waiting wives and children.

At that same moment, in South Wellington, Ladysmith, and Extension, the militia was pounding on doors, rousting the inhabitants from their beds and herding sleep-befuddled men, women, and children into the streets.

> Well I can tell you they arrested the whole Bowater family. There were three scabs with a unit of militia. Now comes the knock on the door; my father went. So it was a big guy and my father asked what he wanted and he said, "Well, I've come to arrest you." My father said, "Please hand me my warrant." And they said, "We don't need one under martial law." So my father got dressed and he went out and the minute he got in the ranks, they said, "You've got a son, haven't you? We want him too." He was sixteen years old, spent his seventeenth birthday in jail. After that they come in and they looked around the house to see if we had any weapons of any kind, but we didn't have any.

Ellen Bowater had been born into a union family and she would marry into another, but her experience that night when she was twenty years old ensured her future as an activist, a "true-blood both union-wise and political-wise" as her nephew would describe her.

> So by this time I'm out in the front yard in my nightgown, never even thought a thing about it. So they go right down the road to Greenwell's and they arrest my brother-in-law. Then they went up the hill where my other two brother-in-laws were and their father and brother and arrested the whole four of them and there was another brother, but they thought he was a girl. At that time they used to sleep in bed with their skull caps and those nightgowns on and they thought my Uncle Don was a woman in bed so they never arrested him.

> My father owed the CPR for the fare for bringing us out from the old country. And we hadn't been there too long so naturally he hadn't paid it back. And they come to the house and told him he had to go to work or they'd put him in jail. "Well," he said, "I'm not going to work." And they didn't put him in jail at that time but after those riots, they come to the house and took my father. And my oldest brother. He would be about eighteen, I guess. And he was in jail for six months. I was thirteen years old and they come to where I was in bed and they tipped the bed up against the wall looking for guns.

The search for a cache of arms, begun the evening before in the Athletic Hall, continued for several days. Highlanders who had been posted near the windows before the soldiers broke in on the meeting "were positive they saw rifles and revolvers being passed from hand to hand inside the hall." But even when they ripped up the floors, they failed to find any weapons. With martial law to remove many of the legal rights that Canadians would normally expect and the militia to provide the muscle, the arrests continued.

> My father was clearing land with my two cousins and they said, "Uncle, aren't you going to the parade today?" The miners were having a parade. So he rushed up there all black, overalls all covered with charcoal, and they paraded to the station, but instead of there being any strikebreakers at the station, it was a bunch of soldiers. So the officer run a cordon up each side

> and closed in all the strikin' miners and the chief of police told 'em which ones to pick out. Supposed to be the ringleaders, see? My father happened to be one of them. I think they took eighteen of them and they put them on the train and sent them to Victoria, not washed or anything. Took them there for three weeks 'til the trial.

> My father was the secretary of the union and he had tried very hard during the strike to keep things under control. One night he dressed up in women's clothes to go to South Wellington to put a mine boss on a boat to Vancouver so the miners would not kill him. When they had a riot at Number One he went down to see what all the noise was about. When the militia arrived, they arrested all the men including my father.

The police arrested strikers and their sympathizers but not one strike-breaker. Although some of the arrests were random, the police took great care to arrest all the local union officials and as many of the international advisers as they could find. Removing leadership is a well-known device to control crowds, but the police went a step further when they arrested MLA John Place.

The charge against Place was possession of a special policeman's revolver. Place had indeed removed Const. Taylor's gun in the incident on the wharf in an effort to defuse what became a very brutal incident. Most of the strikers believed Place had been arrested because he was an MLA. He and Parker Williams had never ceased their demands for government intervention in the strike. When Williams proved impossible to arrest because he had been tending his six children during the riots while his wife was sick, they arrested his son, David, instead.

> I was on a load of oats coming up from the clearing when I saw the car, I believe with three police in it, drive to the house. Dad wanted me to take to the bush, but I told him there was nothing to worry about. When we got to the house, mother was in the backyard with a shotgun. She was preparing to shoot them for having the audacity to arrest her David. I calmed her down and assured her no harm would befall her David and she felt better. I was taken to Ladysmith, joined the rest of the sixty-five and went to Nanaimo on the E & N evening train. In Nanaimo, instead of marching in a direct line from the station to Newcastle Townsite, we were paraded through town, down Bastion Street to Commercial then north past the Plaza and the old post office.

As the police marched the arrested Ladysmith men through Nanaimo's downtown streets, a newcomer was checking into the Wilson Hotel under an assumed name. To anyone who asked, he said he was "one of a group of campers from Shawnigan Lake who had come to Nanaimo to 'see the fun.'" To Attorney General Bowser, who had chosen him particularly because he was "a roughneck who could get in among the mob of men and yell and throw a stone occasionally in order to be one of them," he was an undercover Pinkerton detective whose mission was to find the person responsible for dynamiting McKinnon's house in Ladysmith.

Pinkerton's was an American detective agency that did a lot of work for Canadian police forces, which seldom had trained detectives of their own. The superintendent of the British Columbia Police, F.S. Hussey, had a "warm relationship" with Pinkerton superintendent Philip K. Ahern in Seattle. Ahern was able to provide a suitable man for the job on a moment's notice.

The suitable roughneck set about blending in. He changed his identity from camper to working man; he moved into a boarding house where a number of manual labourers were staying; he bought rounds at the local saloons and was generous with the bottle of bootleg booze he kept on hand; he listened when booze-loosened tongues revealed the time and place of union meetings and news of dissension among members of the union executive. He soon realized that he was not the only spy in town; there were union "spotters" spying on other union men to see if they were faithful to the union. He reported everything he saw to the police and had opinions on many things including the behavior of the strikers' wives.

> I note that the wives of the miners are the most outspoken and bitterest of enemies to the soldiers. They are often cautioned by their men folks when speaking in reckless tones of the soldiers.
>
> Seldom does it fall to the lot of a person to mingle with such a vicious . . . and misguided lot of human beings of that sex; the strings of oaths and foul language which eminated [sic] from the mouths of these mothers and wives would shame a mule driver.

Anne Bryant, whose father was the secretary of the union, bore her anger and sorrow in more genteel fashion.

> So we had to visit my father in jail. It was very humiliating for my family. I was a young wife then and expecting a baby. I went into labour due to the stress. I had to go to a friend's house not far away, through crowds of men carrying guns and my baby was called a strike baby. There were two children born during the riots.

The women, whether quiet or foul-mouthed, were witnessing the arrest and imprisonment of their husbands and fathers. The police had arrested 179 Nanaimo men and soon the Nanaimo jail would hold no more. When the police decided to relieve the crowding by sending forty-five men to a Victoria jail and others to the Coquitlam farm colony on the mainland, the police and their prisoners had to pass between lines of "crying, cursing and hysterical amazons."

David Williams, as his father's surrogate, recorded the process of entry to the jail. Each man was stripped, weighed, measured, fingerprinted, relieved of his tobacco, and confined with two other men to a two-by-three-metre cell with a stack of bunks and a bucket for a toilet.

> At first visitors were allowed to shake hands to pass money to the prisoners, because this was to bribe some of the guards to buy sugar. Soon no contact was allowed. All mail was opened. We were allowed to write one letter a week, one side of one sheet, left open. We were given magazines to read. The blank paper in these was used to write letters, other blank spaces for envelopes and porridge was glue. Personally I was a guest of King George for fifty-two days.

Each prisoner had an enamel cup and plate and a tablespoon. Breakfast was one third of a loaf of bread, black coffee, and rolled oats porridge with no sugar, milk, or salt. Lunch was one third of a loaf of bread, some well-boiled beef, some vegetables and a cup of water. Dinner was one third of a loaf of bread, porridge, and black tea. The porridge contained tangles of jute from the rolled oat sacks. The prisoners were always hungry.

When police had first rounded up the Extension men, they had placed them in a fenced compound. Unable to get food to them legitimately, Annie Gilmour had tossed apples defiantly over the fence to her husband and her son. The brawny Scots woman even stuck an apple on a guard's bayonet to show she wasn't afraid of him. When the soldiers brought their prisoners to trial in Ladysmith, women and children waited outside with food parcels for their men. As the women thrust the food forward, the guards knocked it out of their hands and trampled it into the road.

> When my father was in jail in Nanaimo we used to go downtown. I don't know where we got the money to hire a team of horses and a buggy but we used to go to Nanaimo every once in a while and see him for a couple of hours and come back again. More or less a family affair. He was glad to see us. I don't remember him behind bars at all. There was a visiting room or something.

> Mother couldn't go to see Dad because she had no money. We bought a few pigs and gathered slop from the town to feed them and then sell them to the butcher. We used to get up there at 4:00 in the morning and go to town to gather slops and then go to school.

> My father-in-law didn't say too much about his stay in New Westminster. Jail's jail. Ye had to eat what ye got. They hadn't done anything he and Sam Guthrie, but they were executives and there was some violence so they had to put somebody in.

By August 21st all the arrested men from Ladysmith, South Wellington, Extension, and Nanaimo had been remanded for trial. The only prisoners granted bail were four sixteen-year-olds, but not including sixteen-year-old William Bowater, Jr., who had been charged with burning down Manager Cunningham's house in Extension. The first preliminary hearing was an emotional one.

> *Nanaimo Free Press*, August 21, 1913
> In the little courtroom sufficient seats could not be found for all the prisoners and several had to stand with their backs against the wall. Silently, the crowd, mostly women, filed into the tiny gallery to watch the proceedings and nod encouragement to their friends below. The solemnity of the occasion seemed to impress itself on the visitors and several of the women began to sob softly when their husbands' names were called out. The pale, drawn features of the children and the tired weary look of the mothers showed plainly that the strain of the past few days had not been felt by the men of the city alone.

Thanks to the redoubtable Mrs. Charles Axelson, the Ladysmith hearing was not so somber. Lempi Guthrie, there to support her own husband, Sam, watched as the court called the large and impressive woman to be the first witness. The prosecuting attorney asked Mrs. Axelson if she had been the ringleader of the unlawful assembly. "No," she said, "but I did join in the singing." "Would you oblige us with a verse or two of the song the strikers were singing?" asked the crown prosecutor. "Why certainly your honour," she replied.

> She simply turned the tables on the lawyer. She had a lovely trained voice and in a short time the whole large audience wholeheartedly joined in. The judge tried in every way to stop them and had great difficulty in restoring order to court. There were laughs and boos and the whole proceeding was turned into an empty farce. Prisoners, witnesses, and spectators burst forth in round after round of applause.

But order was soon restored and then the tension began to build as one, then another of the accused, many of them young and unsophisticated in such matters, opted for trial by judge alone rather than by judge and jury. On the advice of their lawyer, who said they would receive a light sentence, the men pleaded guilty to the charges of rioting and unlawful assembly.

> The outcome of the trial was that four of the strike leaders including Sam were sentenced to two years in the federal penitentiary. About thirty others were sentenced to eighteen months at hard labour in Oakalla, the provincial prison. Feelings were running high at the time of the trials which accounted for the very stiff sentences. In the other strike centres, the arrested miners were tried by jury and received much lighter sentences – quite a few were acquitted.

The Extension strikers remained in Victoria until September 12th, when a train brought them back to Nanaimo for their hearing. Friends and relatives packed the station and strained for a glimpse of fathers and husbands they had not seen for weeks. The prisoners came off the train handcuffed together, bearded but healthy. As if to hearten their loved ones, they sang as they passed through the crowds down to the courthouse escorted by soldiers. Among them were men whose names still conjure up images of the strike and the union – Greenwell, Bowater, Gilmour. Police spotted three union officials in the crowd and arrested them for vagrancy. In a decision rendered in far off London England, picketing, a legal right in English law, had that day been declared illegal on Vancouver Island.

The majority of the soldiers had gone home by the end of August, leaving 265 of the original 1000 to guard the five coal towns. With the preliminary hearings over and the troublemakers safely in jail, the residents adjusted gradually to the presence of the militia. Early morning reveille and the intermittent skirl of the pipes imposed themselves on the atmosphere. In the military camps, idle soldiers whitewashed stones and used them to outline pathways and cookhouses. Colour parties hoisted Union Jacks every morning and lowered them every evening.

In Cumberland, the remaining detachment of soldiers had gone to great lengths to make their numbers appear greater than they were. There was much marching and band music. When the order came to send the machine gun to Nanaimo, the horse that had pulled it remained behind and was brought out for daily exercise to give the impression that the gun was still in Cumberland. When the HMS *Shearwater* docked in Comox, twenty-five sailors marched through Cumberland bayonets fixed and bagpipes wailing.

The Pinkerton undercover man was amused by the amount of "noise and display" and disruption of traffic that occurred every time the "kilties" wearing their gorgeous golden-braid-and-ribbon-bedecked uniforms and massive black fur hats, paraded in Nanaimo. Included in his daily jottings for his spymasters was crucial information regarding the undergarments of men wearing kilts.

> It is a positive fact that these men wear no covering over their upper legs and hips other than kilts. They wear a narrow strip of light coloured cloth hung from the front of the belt to the rear and fitting tight over their privates. That is all.

The undercover man revealed a prudish side when he went on to complain that because the Dallas Square camp was elevated on a slope, it was possible for people passing by to view these mysterious undergarments if a soldier happened to forget himself when he was stooping over to tighten a peg or "lying around in a careless manner."

> [T]heir skirts go up so far behind that all passers by, women, children, and the people seated opposite are given a display that one is not supposed to

meet with in a civilized land. I myself have seen some shocking displays there on many occasions.

And the soldiers were often on view. Spectators gathered to watch football games and to observe the soldiers, stripped to the waist, scrubbing themselves and their clothes at long tables set up in the open. It was rumoured that the khaki apron worn over the kilties' sporrans had been added to the Highlander uniform during the Boer War, twelve years before, when enemy sharpshooters had used the sporran as a convenient target.

The militia had become a familiar sight and a welcome one to anyone not on the union side of the strike. Strikebreakers felt safer; the danger of another riot was virtually gone; and local businesses, after months of reduced sales, now had new patrons for their wares.

The militia's aim was to cultivate good will, and if the strenuous public relations gestures also happened to keep the militiamen occupied, all the better. On August 22nd the kilties had held a sports afternoon with a band playing to welcome the public. Two days later four hundred soldiers and sixty specials had attended a church service on the grounds of the courthouse. That evening pipers and drummers had performed on the waterfront.

After the majority of the militia had gone and a more permanent camp with wooden floors for the tents had been set up at the corner of Milton and Wentworth, away from the centre of town, the militia was not as apparent to the average citizen. But the strikers could not forget the soldiers, especially when work resumed at Number One mine in September and armed troops escorted strikebreakers to and from work.

> Men can't fight against guns; there's no use o' trying. You gotta decide what you're gonna do and if you get a gun yourself and kill somebody, well you're – in those days they'd hang you. Today they give you coffee and donuts. But in those days they'd hang you.

The attitude of the union men towards the militia is best expressed in songs and bits of doggerel that came into circulation. Everyone knew the words to the most popular song: "Bowser's Seventy-Twa."

> Oh, did you see the kilties boys?
> The laugh would nearly kilt you boys,
> The day they came to kill both great and sma,
> With bayonet, shot and shell,
> To blow you all to hell,
> Did Bowser and his gallant Seventy-twa.
>
> Then hurrah boys, hurrah,
> For Bowser's Seventy-twa
> The handy, candy, dandy Seventy-twa!

’Twill make the world look sma,
Run on by Colonel Ha
And Bowser and his gallant Seventy-twa.

But despite the animosity, peace prevailed. So peaceful was it that Col. Hall had to plan diversions to keep his men occupied. One such diversion was a mock battle. He divided the militia into the defenders and the invaders. The invaders were to enter the exhibition grounds and the defenders were to stop them anyway they could. Henry Wilby was a twelve-year-old school cadet who came up to Victoria on the E & N to join in.

> We got off the train at the Nanaimo River and found the invaders camped in a field near the railroad bridge. We were immediately attached to the company with the rank of scouts. That night we slept on the ground in the open. Next day we marched and rested and marched for miles and miles along country roads, never on main roads. It took us all day to get to Nanaimo. We weren’t in a hurry. We met a farmer with a load of strawberries on his way to town to sell them. He was soon persuaded to sell them right there. It was not long after that when we met a wagon full of bottled beer. The beer went the same way as the strawberries and called for another rest by the wayside. Oh, and we also won the battle by tricking the opposition and never firing a shot.

War games kept twelve-year-old cadets amused and grown-up soldiers occupied. But these soldiers were militiamen used to putting on their uniforms for only two weeks a year not for months on end. A dollar a day plus room and board did not compare favourably with the $18 to $25 a week most would have been receiving at their regular jobs. Employers in Victoria and Vancouver were threatening dismissal unless absent employees returned quickly.

At the end of October, Bowser announced that 75 percent of those remaining would go because the government would pay the cost no longer. The mayors were horrified. They could foresee a return of the violence as more and more union men returned to work. Col. Hall agreed and he won the day. The 265 militiamen remained until the next summer.

But now the militia was there because the towns themselves – not the provincial government – wanted it to be. Soldiers continued to escort strikebreakers and supervise payday. There were incidents between small groups of strikebreakers and strikers all through the winter and into the spring. In July 1914 there were reports of miners carrying guns. Violence could have erupted again on Vancouver Island if it had not been for the violence of a much greater magnitude brewing in Europe.

For the men and their families involved in the 1913 riots, however, the machinations of European politicians were unimportant in relation to what they faced beginning in October 1913 when the strikers’ trials began. Judge Frederick William Howay, friend of the powerful and amateur historian of whom it was said that he was "far more competent as an historian than a judge," arrived in Nanaimo to conduct the so-called "Speedy Trial" of the Ladysmith men. The label derived from the fact that some of the defendants pled guilty in return for a reduced sentence. When the crown added additional charges at the beginning of the trial, all but one of the defendants changed his plea to not guilty.

This trial is famous, however, because of the judge's final statement before sentencing. After several days of evidence to establish who was in the mobs that pushed and yelled and marched and hurled stones in those eventful hours in Ladysmith on August 11th and 12th, and after a defense presentation designed to make public the legitimate grievances of the strikers, the judge announced on October 23rd to a courtroom crammed with relatives of the accused that this "out of the ordinary" trial required him to go beyond the usual brief comment made by a judge before sentencing.

Howay was of less than average height and had a carefully coifed curl on his forehead, but his words had a severity that belied his cherubic appearance. Having assured the men that they had been well defended, he proceeded.

> Your counsel has said all that could be said and if I have found you guilty, it is not because your counsel has erred but because you have woven around you a net of circumstances beyond the power of any counsel to untangle. This was not an ordinary riot, it was not a sudden ebullition of pent-up feeling, but shows all down the line a deliberate scheme, a design from one end to the other. The riots in Nanaimo, South Wellington, Extension, and Ladysmith were all for one purpose, were simultaneous and carried out with one line of action. Bombs were thrown, property destroyed and peaceful citizens made to flee for their lives in the persistent state of terrorism indulged in.

Thus, although no such premeditation had been proven or even mentioned during the trial, the judge included it in his deliberations along with the defendants from South Wellington, Extension, and Nanaimo, who had not even appeared before him. Then the judge described the riot in detail: the bomb throwing at the Temperance Hotel and the McKinnon house, the mobs marching around the town and the stoning of certain houses.

> I have looked over your faces to see if I could see any sign of sorrow or repentance for what you have done but I fail to find one man among you to express sorrow for his lawless acts. Your counsel knows there is no more sympathetic man than myself – one ever ready to extend mercy – but I have read over the depositions and find little mercy you have shown . . . I was appealed to on behalf of your wives and children but what do I find here? I find your

women singing "Drive the Scabs Away" and throwing rocks themselves and these actions take away very much of the strength of the appeal for mercy on your behalf, because of your women.

Having heaped the crimes of the wives upon the heads of the husbands, Howay then gave sentence. Five men including Sam Guthrie and Joe Taylor received the maximum penalty for rioting: two years. Twenty-two men including Charles Axelson received one year and a fine of $100. Eleven were sentenced to three months and a fine of $50. In the opinion of the crown prosecutor, the severity of the sentences would lead to a "dissolution of the forces of agitation." The son of one of those sinister forces gave this picture of his father's trial.

There were six of us kids at that time and my father was imprisoned for eight months. He wasn't educated and could not speak so good 'cause he came from Yugoslavia. He could understand everything but couldn't express himself. Anyway he was brought up before the judge and the judge says, "Guilt or not guilty." My father says, "I'll leave it to you judge." The judge sent him for one year to Oakalla. My father says, "Thank you very much" and the whole courtroom was laughing.

After this trial the court moved to New Westminster, the Nanaimo area having been deemed too hostile, not to the defendants as is usually the case, but to the judge, who further damaged judicial reputation for impartiality by giving an interview to the *Westminster News* showing the miners in an unfavourable light.

The "Strikers' Assizes," the trials of the Nanaimo, Cumberland, South Wellington, and Extension men who had been in jail since August, began in November and lasted until March 27, 1914, when 114 jurors received their pay and discharge. Judge Aulay Morrison had replaced Howay. He had a reputation for courage and strength of character and as a man who spoke in favour of unions but not, of course, of violence.

And I remember when they were up in front of the judge and he said, "The finest looking body of men I ever had in front of me. But," he said, "you know we have to uphold the law."

I ran across the Judge a little while ago when Nuenthal appeared in court to give evidence and he told me that afternoon that he was deeply touched by Nuenthal's appearance and haunted look and felt tempted to let him off then and there.

Of the 213 men arrested, 166 were actually tried and of these only 50 – including 38 from Ladysmith – received jail sentences. Whether it was due to the change in venue, the change in judge, the increased time elapsed between the riot and the New Westminster trials or a combination of these factors, most of the defendants at the Strikers' Assizes went free. All six of the Cumberland men were acquitted, seven of the ten South

Wellington men went free and thirty-three Nanaimo men pled guilty, but none of them went to jail.

The police had charged several Extension men with more serious offences such as arson and attempted murder, but only Joe Angelo and William Bowater, Jr. were refused bail when it was finally granted in November. Sixteen-year-old Bowater sat in jail for seven months until his trial at which the jury was unable to reach a verdict. A second trial was dismissed. He had been allowed out of jail for only one day in that time – Christmas Day.

> Uncle Bill Bowater was supposed to have set fire to the company boss' house. He was released for Christmas 1913 on $50,000 bail. Half had to be in cash and half in land securities. The fifty thousand was raised. I know my uncle put up his farm and miners put up their property and I just forget how the cash was raised, but it was just like raisin' a million dollars today.

On March 18, 1914, thirty-three Nanaimo men pled guilty to unlawful assembly in return for the withdrawal of twenty-three other charges. A large crowd gathered on Johnson's wharf to welcome them home on their return to Nanaimo on board the S.S. *Cowichan.* They rode high on the shoulders of their friends, through a cheering crowd of hundreds singing the "Marseillaise." Four of the celebrants ended up in jail charged with being drunk and leading the crowd in singing. Four days later, the UMWA held a rally at the Athletic Arena and collected $46.25 for their fines. Speakers at the rally cautioned the thirty-three on suspended sentences to be very careful.

Twenty-three Ladysmith had been imprisoned at the Oakalla prison farm, but only twenty-two came out on April 3, 1914. Joseph Mairs, Jr., was not with them. He had had tuberculosis before he went to prison, and though he was sick, he had been assigned to clear land. With no doctor in residence or medical examinations available for prisoners, he sickened and died two weeks short of his twenty-second birthday to become the martyr for the union cause. The marker stone on his grave in Ladysmith, the largest stone there, was bought with money raised from the sale of postcards showing Mairs with his racing bicycle.

> Remember me as you pass by,
> As you are now, so once was I,
> As I am now soon you will be
> Prepare for death, to follow me.
>
> A martyr to a noble cause,
> The emancipation of his fellow man.

The amnesty arrangement that came too late for Joseph Mairs was set in motion in January 1914 when sixty wives and mothers of the imprisoned miners visited Premier McBride in Victoria. Their dignified petition contrasts with the threat made by the Miners' Liberation League to use a general strike to free their brethren. Formed in October 1913 by the Socialist Party of Canada, the Social Democratic Party, The Industrial Workers of the World (IWW), and the Trades and Labor Congress, and led by the wild and intemperate Robert Gosden of the IWW, the League held rallies to protest the federal and provincial governments' handling of the strike and the harshness of the prison sentences. On the streets of Nanaimo, Vancouver, Victoria, and New Westminster, members and their wives sold League buttons and display cards for store windows showing a prisoner dressed in stripes while his wife and two children stood outside his cell.

But the League was an uncomfortable alliance of radicals and conservative unionists. When Gosden threatened Premier McBride in a speech made in December 1913 by saying, "If McBride or any of his gang go hunting they will be foolish, or if they are wise they will get some 'sucker' to taste their coffee before they drink it," Frank Farrington withdrew the UMWA.

No amount of petitions or threats altered the strike situation, however. Mine owners, summoned before McBride on March 17, 1914, declared themselves quite happy with things as they were. Production figures were climbing; new mines like Number Eight in Cumberland and Reserve and Morden mines near Nanaimo were in the final stages of development. Six hundred fifty men, half the normal work force, were working for Western Fuel, one hundred for PCC. And with markets depressed and the Cumberland mines on short time, the last thing the mine owners needed was more employees. Let the strike go on. The owners could resist the UMWA forever.

It was difficult for the strikers to hold out. The UMWA asked members from all across North America to contribute 50 cents to shore up the Vancouver Island strike fund, which would soon run out, and the union attempted to stiffen a few backbones with a proposal to publish the names of every union man who went back to work. In his regular written report, the Pinkerton operative expressed sympathy for the strikers as he noted the presence of more outsiders in town.

> The new miners now being employed are of the lower European class, laborers mostly and ones who will never patronize any business house or good hotel such as the town furnishes to the unmarried class of miners heretofore. The new arrivals are the sort that huddle together in rooms and hovels, and never patronize anyone outside their own nationality and on payday send most of their earnings home to the old country.

To go back to work after so many months was the act of desperate men. In the spring of 1914 there were many desperate men.

> Some had worked all along because they thought they were right on account of their religious beliefs, but a lot went back to work because of the money. They were hard up and had to do it to feed their kids and stuff. Actually, the ones that scabbed didn't suffer as much as what their kids did especially in a place like Extension where there were pretty strong union people.

> My dad was a fire boss and he worked all through the strike. And he was frowned upon. All he was doing down there was keeping the pumps going to keep the water out because the mines were under water, so that when the strike was over, the men would have a job to go back to. And we were going to school at the time and a lot of the school kids had no use for us.

> I guess a lot of people were beginning to get down to putting in their short laces. There was a lot of hard feeling in families. Some members of the family wanted to go back to work and other members of the same family said, "No, you're not going back to work." So to this day there's families in this area that don't talk to each other.

> I finally asked my father about it and he said to me, "I expected you to ask that question. Well," he says, "when I went and asked for my job back there was over a thousand men working. The strike was lost anyway."

Newspapers reported incidents almost daily of miners being molested by strikers. Strikebreakers' wives were said to be gathering in one house for protection while the men were at work. They kept their children home from school. Three hundred extra militiamen returned to Nanaimo to reinforce the Mayor's strict regulations for conduct on May Day: all saloons closed, no demonstrations within city limits, no loitering, no public holiday, every picnicker searched, and no parade to the Cricket Grounds.

But the union folk still gathered at the Cricket Grounds, which was outside the city limits and therefore outside Mayor Shaw's domain. And as always, the high point of the day for the adults was the speeches. Robert Foster walked to the podium to the singing of "Death of Nelson" with new words that began, "Now boys lay down your tools..." The union president told the crowd of the total lack of success in opening up negotiations with the owners, but promised "Vancouver Island coal mines will be organized no matter how long it may take to carry out the work of organization and in spite of any temporary delay."

On June 25, 1914, union locals met to consider a set of proposals by the owners, who had finally bowed to pressure from the premier. But the owners had refused to include recognition of the union in their proposal. The vote was 1167 to 274 against acceptance.

The British Columbia Federation of Labour (BCFL) had been talking about a general protest strike since September 1913, but it wasn't until the following July that enough members agreed to even discuss it at a special convention. The UMWA was against the idea because it would break agreements already signed. Experience had taught the union that general strikes only cemented public hatred for the union movement. But the convention voted in favour of a general strike referendum to be held on August 18th.

On opening day of the convention, Robert Foster had delivered a fiery speech full of vituperation against "lickspittles of strikebreakers" and "scab herders." One week later, a much more subdued Foster announced that there was no more strike pay for Island miners. Having poured $1.25 million into the Island and having to support striking miners in West Virginia, Iowa, and Colorado, the UMWA could only offer a $15,000 fund for special needs. It was July 21st.

Two weeks later, Austria-Hungary and Germany declared war on Russia, France, Britain, and Serbia, and the big Vancouver Island coal strike fizzled out. The soldiers departed for the battlefields of Europe. The date for the BCFL referendum passed unnoticed. On August 20th, the dock where angry mobs had jostled specials and mocked strikebreakers was now the scene of tearful goodbyes as Nanaimo's army reservists took leave of their families. On the same day, striking miners accepted the company proposals they had rejected in June. All former employees would be allowed to return and to maintain their membership in the UMWA. But the companies would not recognize the union.

The only bright spot in the settlement was the release of the men who had been in jail. Sam Guthrie came home to his wife and a baby son born while he was in jail. Twenty years later, when he had been a much-admired MLA for many years, Sam Guthrie was driving by the penitentiary with a friend. "I cried myself to sleep here on my bunk night after night," he said in his Scottish burr, "wondering what was to become of me and my family."

> I remember when they got out of jail. We knew they were coming. We went down to Chase River and I run all the way back holding my brother's hand. I was so pleased to see him.

> From then on it was kind of chaos you know. The miners stopped getting money. I guess the union spent all they could and some men went away, all kinds went away, pulled out, and got jobs in other places. That was at the end of the strike. Course you couldn't get much work and we had to live off the country again. But it was worth it. Absolutely.

> Well certainly it was worth it. Because the coal miners were the least paid on the Island. They had low wages for dangerous work.

I don't think there's anybody what thinks the strike was carried on far longer than there was any sense. Because the company beat them.

As if to underline the futility of the two-year battle, an event had occurred on August 11th that made the future even more precarious for the people of South Wellington. Looking back, many miners associate the fire that happened that day with the riots of the year before. The image of flames leaping from one house to the next to the next seems to match with the memory of men cursing and beating other men as they fled into the forest. In reality, the bush fire that burned one whole side of the main street and then, fed by a complete change in the direction of the wind, roared back down the other side, came when the strike was all but over. Just when the strikers were about to resume work, the fire destroyed their homes.

The newspapers blamed it on a bush fire started by accident and encouraged by the hot dry weather and the brisk wind. Not everyone agreed.

Well, apparently in South Wellington there was just the one main street through it and there were houses on both sides. And on one side the union people lived and on the other side scab people lived. So on this day the wind was blowing along, so the union guys lit the end house of the scabs and it went right along and cleaned out all their houses and when it got to the end, it crossed over the street and the wind changed and it came right back through and burnt all the union guys' houses out too.

It was always suspected that some of the strikers started that fire. Whether that was someone's guess or not, I don't know. I know I had nothing to do with it.

The South Wellington fire had nothing to do with the strike. The big fire went through there and the hotel and the whole darn outfit was all burnt. Only the hall was left standin' and one house down here. Caused by a bush fire. Yep.

Bush fire or act of retribution, the South Wellington fire made a tragic closing act to the strike. But the effects of the strike would be felt for years to come. Blacklisting, that most effective way of punishing workers, followed quickly. The owners with their eyes on the war and declining markets and their fear of unions undiminished, reneged on the agreement made on August 20th. Union men could not get work as miners on Vancouver Island.

My dad was on his way back to work and one of the bosses had just come up to the surface and he yells, "Hey, John Cottle. Where are you going? Don't you know you're blacklisted in the Island?" And my father found out it was true too.

> If you were blacklisted from one mine, say Canadian Collieries, they owned about six mines so you were blacklisted from those six. If you were blacklisted by Western Fuel, you were out of about five mines. Morden was a neutral company and so was Jingle Pot. And there was a tendency for co-operation between companies. They all peed in the same pot so to say.

> If you were blacklisted in Nanaimo, you was also blacklisted in Cumberland, and you might as well get off the Island altogether. No protection at all. And that's all we were fighting for.

Faced with a continuation of their desperate situation, the miners fell back on their legendary resourcefulness. Single men could join the army or navy or find a coal mine somewhere else. Married men looked for non-mining jobs or did relief work, and continued to provide food in the same way they had during the strike.

> During the strike there was a lot of single men and they stayed for about a year, year and a half and when there was still no sign of it ending they took off for the States. Some went to Montana, one of my cousins went to Colorado. But the married men, they couldn't. They had their families so they stuck it out.

> It was a very, very bad thing in Nanaimo. Many families immigrated back to the old country as soon as they got money from the Salvation Army. Others stuck it out and nearly starved to death.

> My father was unemployed for eighteen months. We made ends meet by growing a garden and living as close as possible. I started to work when I was thirteen peddling milk. My salary was $15 a month and it went into the pot.

> My father was blackballed for about two years before he could get back in the mine. They had a relief gang then, so he worked on the road for the government, putting roads in all day, and he used to get so much for relief per month.

> After the Big Strike, father was kep' out of the mines for two years on account of his activities. And when the Reserve explosion took place in 1915, he was working for Kaiser at the time paving Haliburton Street and they sent him to go out on the rescue team. And the old man tells the story that after the explosion he got a cheque from the company and he says, "Their books would never balance 'cause that cheque was never cashed."

> There was a mine near Lethbridge and they sent a man over here 'cause they knew there were lots of good coal miners here and they collected a whole bunch of these old strikers and they all went to this mine at Kip. In fact, I went down there myself. And we stayed one winter. It was cold down there in Alberta you know.

> So eventually Dad went up to Fernie. He was told of good work up there. But even up there he just worked on day wages and doing little odd jobs. He

said the country was rough and cold in the winter. My mother wasn't in the best of health so he said he would never take us up there. So we stayed at Rock City while he was away.

There was no welfare, my father's health broke down and so that made me the provider. I was anxious to get out of school and get workin'.

Because the immediate effect of the war on mining had been a panic closure of all metalliferous mines and particularly copper mines, these closures decreased the need for the coal and coke used in copper smelters and for mine locomotives. But by 1915, the demand for copper for the war effort increased the size of the market for coal and coke as well. Production climbed and with it the demand for expert miners. Miners who had enlisted could come home if they wished and blacklisted miners were able to get mining jobs wherever they wanted.

That's right. That's how I got out of the navy. So we got the opportunity, anyone that wished to go back to the mine, they'd get an honourable discharge. The ship I was on, HMS *Galiano*, went down after that. I think there was about twelve on board that ship that was ex-miners and I think there was about six of us left for the mines. The other six stayed with it and went down with the ship.

After the South Wellington flood, there were so many killed and so many scared, they were needin' experienced miners. But the superintendent says to my dad, "I can't give you a job 'cause you're blacklisted. If your name goes through the office, you'll be laid off." And my father says, "Even a few days would help." So he started him and the superintendent says, "I'll tell you what we'll do. Move into one of the company houses." So we did. We moved from Extension and the word came through that they wanted to lay him off. "Well," the superintendent said, "he's got so many children, we can't get him out of the house." And they needed miners. They needed coal for the war so they left him alone.

My mother was bathing us kids in the kitchen just before Christmas. My dad was blacklisted and she got to thinking about Christmas and having no money I guess and she started to cry. Just then Dad walked in carrying his tallies and a new pick. He was working again.

With local miners returning to work, the company lost interest in the strikebreakers they had lured to the Island during the strike.

I felt sorry for the scabbies. Guys from the prairies would come out to work during a strike, they'd break a strike and as soon as it was over they'd get rid of them. They didn't want 'em ya know and they wouldn't work 'em and that's all the thanks they got. They only wanted them in there to break the union.

All the scabs that was livin' down at the mine, they just fired them. Just like nothin'. There was only two that stayed on. Union men was better men I guess.

> There was enough scabs in Cumberland to run the mines partially. And you had immigrants besides at that time that had immigrated let's say from mostly foreign countries what split up when the war broke out. They didn't realize what they were coming to so they were totally caught in the middle.

For the local men who were strikebreakers, a special kind of hell evolved. Hatred seldom flared openly as the years went by but the people never forgot. As generation followed generation, the ones who were strikebreakers were always known and remembered.

> There was a lot of bitterness. Even forty or fifty years after. I guess people were still talking to each other, but deep down I guess they wouldn't be friendly. I guess they won't forget those days. They had a tough time to be out of work for that long.

> There was hate in that strike. There was hate. Friends all their life, some worked, some didn't. And the one that didn't work never spoke to them for the rest of his life. And even to this day, one fellow there he come down to South Wellington and he wanted to go hunting with us. "Well," I told him, "you scabbed. None of the guys want you to go." I knew what happened how he had to work and he had pressure on him you know. If a man's wife isn't with him it makes him pretty miserable.

> There were families split down the middle. The father and some of the brothers went to work, some of the other brothers didn't work. And the feeling went down from one generation to the next one. I can remember as we grew older, the families were pointed out to us and what they did in the 1912 strike.

> Well I know one family where two brothers lived in the same room and one of them scabbed and one didn't. And one guy had a curtain down the middle of the room and he put "scab" on one side and "union man" on the other.

> Some people were very religious and they really thought that they were going by the Bible. They thought they were doing right going back to work. It even split up some churches.

> It was sad because the Plymouth Brethren meetings had both sides there. And the ones that were working they thought they were doing right according to what they could see out of the Bible and the ones that wouldn't work went on strike and they thought they were doing right too.

To have been a strikebreaker marked a man for life. Even the most dedicated union man will admit that there were many extenuating circumstances that drove men back to work, but strikebreaking still had to be lived down. Some of the most dedicated workers in the revival of the union in the 1930s were the sons of men who had been strikebreakers.

In the defeat of the union in 1914 were the seeds for its revival. The very failure to gain union recognition provided the necessary rage and determination to try again. And there was another positive product of the strike. The years of making do and helping each other also nurtured a co-operative spirit in Nanaimo. Although the inability of people to pay their bills during the strike had caused the co-operative store in Nanaimo – established before the turn of the twentieth century – to close in the 1920s, the co-operative impulse remained alive and bloomed again.

> Some families had miners and farmers and a lot of times those that came out to farm were related to those that came out to mine. And so they helped each other out. So I guess this helped the co-op, well the miners getting credit from the co-op and the farmers giving the miners food and the farmers helping the co-op to finance the miners in their credit.

> After the strike my father couldn't get work in the mines because of his activities so we moved to Nanaimo and he worked for the co-op. It was badly crippled financially during the strike because of the credit they gave the miners and one of the pioneer families that helped it out considerably was Mrs. Fiddick.

The war was on and the miners were back at work, but the future of coal as the chief source of the world's energy had been threatened even before the strike. Between 1912 and 1914, the mines were able to keep up with demand even with a smaller labour force because the market was so depressed. This was partly due to the fact that consumers had been forced to find other sources of energy when the supply from Vancouver Island slowed to a trickle in the fall of 1912. The artificial boom in copper during the war collapsed after the war. Oil would take a larger and larger share of the energy market. The bitter two-year strike began to look like nothing more than a last pathetic battle before all the combatants succumbed.

The combatants were locked in battle over the right of the men to dignity and security and the right of the owners to run their operations without outside interference. The government wanted to preserve order and capitalism; the union wanted to build a powerful international bargaining tool. Each party could argue the rightness of its goals but it is hard to ignore the fact that the union, the companies, and the government were disproportionately powerful compared to the miners. These men with their legitimate grievances turned to the union because the government and the owners had already failed them. In 1914 the union failed them too.

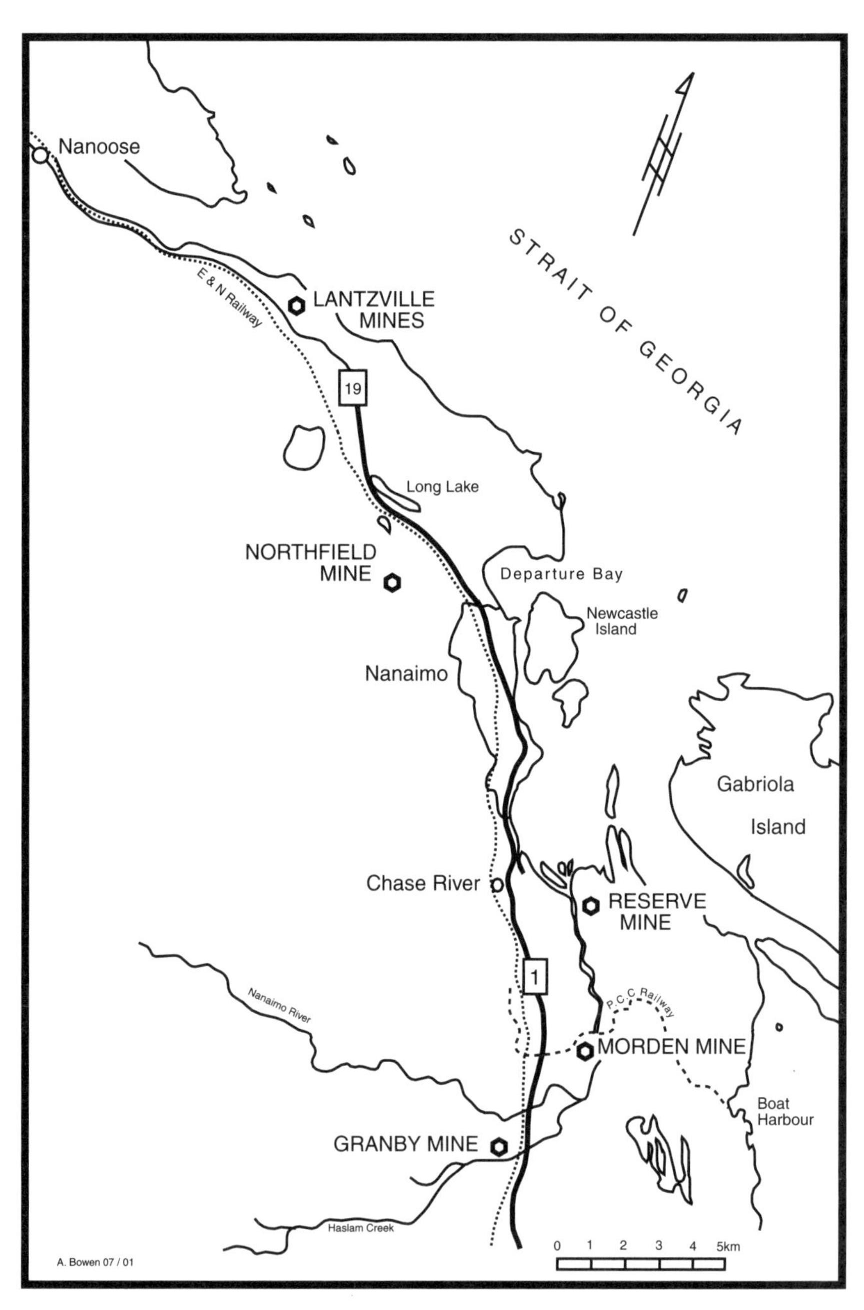

The Post-Strike Mines of the Nanaimo Area.

Chapter IX

Haulin' Coal and Arguin'

My family used to camp at Departure Bay in a tent each summer and Dad worked at Brechin mine. The summer of 1914 there was a party on the Loudon property at Wellington and a group of us walked up from Departure Bay. They had chairs and tables set up on this huge lawn around their huge gray house. All of a sudden someone came running out of the house and he said, "We're at war. We're at war with Germany." It was August 4, 1914 – Rhoda Beck

I was twenty years old in 1914 and I couldn't get my job back in the mine because of the blacklisting so I joined the Seaforth Highlanders. They whisked us right off to war, called us the Canadian Scottish. Before I knew it I was at the battle of Ypres in Belgium and I hadn't even been to training camp yet. I was only there three weeks when I was shot in the head. Spent seven weeks in a hospital in England, got sent back to Victoria, they put a bone graft in my head, and by 1915 I was back in Nanaimo, driver boss in Reserve mine. – Alex Menzies

Just when coal markets began to revive in response to the needs of the war machine and the mines were in need of more labour, 115 enemy aliens of German and Austrian descent were interned, leaving even more vacancies in Vancouver Island mines. When Bill Bowser asked the coal companies to rehire unemployed miners, the attorney general may have been trying to boost wartime production, but he was also anxious to get these men, especially those with families, off the relief rolls. In this matter, if in no other, the miners agreed with Bowser.

Some miners had avoided the relief rolls by finding work at the Morden mine, a new and unusual facility built by PCC during the strike. On a site south of Nanaimo and east of South Wellington, the company had used the idle time to build a pithead entirely of reinforced concrete and steel. The new structure was absolutely fireproof. But the fancy new concrete tipple with its unique flat hoisting rope, its automatic method of dumping cars, and its well-constructed shaft, did not ensure the mine's success or make it a comfortable place to work.

It wasn't such a great mine. It was terrifically gassy. For the first couple of hours you went down there you'd be absolutely sick until I guess your blood stream got the gas mixed in and you were back on your feet again. My dad worked an hour and a half there and he couldn't take it, probably because he had bronchial asthma. My brother lasted seven hours and was working so slow that they fired him. He never got paid for those seven hours.

By 1922 Morden and her sister mine at Suquash on the northern end of Vancouver Island had closed. A new company, Canadian Coal and Iron Company Ltd., attempted to pump the water out of the flooded Morden workings in 1930, but large cave-ins on the main slope forced abandonment of the mine. Generations later, the Morden pithead is the only one still standing, its uncharacteristic structure a strange commemoration of Island coal mines.

In contrast to the short-lived Morden mine, the new Western Fuel Company Reserve mine, so named because it occupied First Nations land at the south end of the Nanaimo Harbour, had a brighter future. The strike had also interrupted the development of this virgin coalfield, but work on the pithead had continued, and the mine was now ready for full production.

The first men to work in Reserve mine remember when the stables were above ground and how beautiful the horses were as they came off the cage and ran free, their tails in the air to celebrate the joy of being in the sunshine again. Even after the stables moved below ground, the Reserve mine animals received the best of care. Living out their lives in the even temperature below ground, their coats never changing in the seasonless atmosphere, they were sleek and well fed. Especially a horse named King, who shared his stall with some chickens. Every day he ate four or five eggs in addition to his helping of oats and hay.

Horse, mule, and miner liked Reserve mine too for its high roof. A coal seam that was as high as seventeen metres in one notable area meant that man and beast could stand tall. But that seam had its problems: It lay in folds and overlaps resembling ocean waves. When the seam pinched down to nothing, "John Hunt's Specials" came in to drill and tunnel through the rock to relocate the seam. They were the rock men, named for the man they worked for, Western Fuel Manager John Hunt. They made fifty cents a day extra for their dangerous work every one of them knowing that inhaled rock dust caused silicosis and early death.

Well I stayed clear from the rock dust. Coal dust comes out, but not rock dust. It's like cement. My father used to say, "John, stay clear of that stuff." A lot of 'em think they're tough. Great big Scotchmen, three hundred pounders, tougher 'n' nails. They all died young. We used to call them John Hunt's Specials. They'd do anything. They're all buried now, every one of them. They'd be all white from the rock dust. In them days you never heard of these respirators or nothin'.

A rock miner who was driving always has silicosis and if you had seen the conditions they worked under . . . The company did the best they could as

> far as getting air into the faces of those places. They drove through straight rock to get to the coal and they drove and drove for eight hours. You couldn't see for maybe 100 yards. They were getting all that in their lungs. They were the big paid men in those days but they paid for it every minute. The rock men they all died young fellas.

Many say that coal dust is just as lethal as rock dust, and everyone in the mine breathed in coal dust. They remember the men who spat black phlegm and died before their time. They cite the example of miners in the U.S. who got compensation for "black lung," albeit after a long struggle. But most miners agree that everyone had black spittle – it meant they were getting the dust out of their lungs before it could do any harm. The majority saw no need to wear protective masks.

> Nobody worried about coal dust. I worked in the mine for years and years and it never bothered anybody. Unless you worked in a rock tunnel, that was different. Miners are a pretty tough, hardy lot. Quite a few of them lived to be seventy or eighty. Any that died before that did on account of their misfortunes.

According to Dr. W.A. McCulloch when he was head of British Columbia's Tuberculosis Chest Clinic, black lung or coal miners' fibrosis is a phenomenon that shows up on autopsy in the lungs of anyone who has lived in a coal mining area. There is no evidence that it causes death or even impaired breathing. But silicosis is a different disease and a much more severe one. The silica in rock dust causes an inflammation of the lung, which makes the miners much more susceptible to tuberculosis and other lung infections.

For years, the WCB refused to recognize silicosis as a separate disease from "black lung." But the high incidence of silica-bearing rock in Vancouver Island mines has gradually caused the Board to recognize that deaths from silicosis-related lung conditions are job-related. One miner well on in years, who attributed his long life to the advice he received from an old German doctor, took matters into his own hands.

> We were working the rock tunnels and he says, "I tell you what you do. You go out to the sea every day and shove your head in the water, breathe that water up your nose and blow it out." And you know it washed that cement out of there, salt water did. And maybe that has something to do with me living longer. I don't know. We used to go swimming in the summer time and my father couldn't swim a stroke, but he'd stick his head down in the salt water and get a mouthful.

Reserve mine had other hazards in addition to rock dust. Sulfur exposure wasn't lethal, but the immediate effects were far more unpleasant. Pithead workers remember blinded miners being led off the double-deck cage. In some areas of Reserve mine, the men ran the risk of being "sulfured" every three or four days.

> I remember in Reserve when I worked on the pithead all these guys coming up with sulfur burns, their eyes and throats all burned so bad someone had to lead them out of the mine. They'd have to sit there until the bus came to take them home. Sometimes they came back the next day but usually they were off for two or three.
>
> You can smell sulfur, but there is only two places it can hit ya. In the stomach and the eyes, but mostly the eyes. If it hits you in the stomach it's bad. When it hits you in the eyes they swell up and water and you think you have an eye full of sand. Only one cure for it. The old country spent millions of dollars trying to find a cure for sulfur, but all you can do is go into a dark room and stay there.
>
> You worked one day; you was off the next. You worked one day and you saw all the colours on the rainbow. You look at a light and you see rings around all colours, everywhere. When we come home we used to get tea leaves and cold water in a rag and put it over our eyes. Cool our eyes off. Went to the doctor's. 'E couldn't do nothin'.

There were many ailments for which the doctor could offer only reassurance and homely advice and let nature take care of the curing. But nature's curative powers were worthless when a man was killed. Early in his practice, Dr. Alan Hall was frustrated and sad when he went to Reserve mine to confirm the death of a young miner.

> They phoned me one morning at the office and said that there's been an accident and they wanted a doctor there right away. When I got out there they were just bringing the man out of the mine on a stretcher. I examined him with all these hardened miners looking down at me, a young fellow, wondering whether I knew very much or not and I felt I didn't know very much at that time. So I gave this fellow a real going over to make sure he was dead and he was, unfortunately. And the thing I will always remember was the manager asking me to stop in and tell his widow. They lived in Chase River. So on the way back to Nanaimo I found the house and went around to the back door and here was a young baby on the porch and its mother was hanging clothes and I had to tell her that her husband was dead. It ended with us both crying.

Reserve mine was not a big killer as mines go; most of the men it killed died years later of lung complications. But there was one notable exception that occurred not long after the strike. On May 27, 1915, the afternoon shift had just gone down the shaft. Reserve was not a particularly gassy mine, but there was always a certain amount of gas in a new mine. Even in the event of a sudden release of gas, however, good ventilation and safety lamps should prevent gas explodions. Reserve mine had both.

Only malignant fate can account for what happened. A sudden outburst of gas pushed a large amount of coal off the face. Only one miner was in its path and he almost escaped, but the fall of coal caught him around the ankles. As he fell to the ground, the glass in his lamp broke and exposed the flame to the atmosphere, an atmosphere laden with gas. The explosion killed twenty-five men.

But because lethal explosions were unusual in Reserve mine, the miners' memories centre around the unusual rather than the deadly. The water shaft was a good example. In 1927, the company sank a second shaft to dewater the mine. Other mines pumped water out, but Reserve hoisted it with two four-tonne cast iron buckets. The water shaft was also an emergency exit when the cage was inoperative.

> Something had broken on the main hoist and they had to send back to Scotland for a part so we had to go up Number Two in the water bucket. To get to Number Two we had to go down to the old workings and the old trails and sometimes we had to crawl and when we got to Number Two shaft bottom there was fungus there just like you were in heaven, hanging twelve feet long. Some of the young boys was scared but anyways, we got out. We had to stand on the edge of the bucket and hang on to the chains to come up. It was wintertime and by the time we got to the top our clothes were frozen solid on us. The boss took us into the office and gave us a shot of whisky and he drove us home and we was all fine. Didn't think nothin' of it.

Market failure closed Reserve from 1930 to 1934. When it reopened, the mine gave jobs to a lot of Depression-idle men, but the work was tunnelling through rock. And there was the problem of fires caused by spontaneous combustion in the old workings and the high incidence of sulfur in the rock areas. The number of illnesses and accidents made further development unwise. After a final fire in December of 1939, Reserve mine closed for good.

Reserve mine's high seam had meant that miners were able to stand upright as they worked. Another post-war mine had the same feature. The two- to seven-metre seam and the far-above-average living conditions drew men to the Granby mine on Haslam Creek nor far from the settlement of Cassidy south of Nanaimo.

On thirty-two hectares of land, the Granby Consolidated Mining and Smelting Company had built a model town. Rows of modern stucco houses lined boulevards; flower gardens, and fruit trees surrounded each house. An athletic park big enough to allow a baseball game and a football match at the same time, the spectators sitting on the natural grandstand of a tree-shaded rise on one side, was only one of the numerous facilities provided for employees. The bachelors' rooming house, with its open verandah and second-storey balcony that gave each room its own access to the outdoors, resembled a Spanish hotel. A resident could dance, read, play pool or billiards, wrestle, or box in the recreation hall. A vine-covered pergola led to a well-insulated dining hall – cool in summer, warm in winter. A full-time doctor, employed by the miners' sickness and accident fund, worked out of a small well-equipped hospital provided by the company.

Granby camp was a marvel when it opened in 1918 and the most amazing thing of all was the washhouse – first in the district. Before there were washhouses, a miner went home black from coal dust and wet from his own sweat and from water seepage in the mine. A twenty-gallon galvanized tub of hot water waited for him at home but home was a long walk or a half-hour train ride or a twenty-minute barge voyage away. The miner's wife or mother would have hauled bucket after bucket of well water into the house to be heated in a boiler on the stove in time to fill the galvanized tub in the middle of the kitchen floor.

In the Granby washhouse there were hot showers, lockers to hold clean clothes, and a drying room heated with steam coils to dry and store work clothes. A man could go home feeling clean leaving his working clothes behind at least until the end of the week when he crammed them into a forty-nine-pound flour sack and brought them home for a woman to wash them.

Extension added a washhouse almost immediately, and though it was not as deluxe as Granby's it was easier for the women to sneak in and use it.

> One of the fire bosses wanted me to go see something with him. We went down to the washhouse and it was full o' women. First of all we seen a woman goin' out and a woman goin' in. Some of them were holdin' their heads past me. They were housewives, daughters. They thought the miners were all in the mine. That's one time they got fooled.

For a long time, South Wellington men went without a washhouse, although some of the pithead men at the PCC mine used to douse themselves with hot water from an outside outlet in the boilers. When Number Ten mine opened in 1938, however, they got their washhouse too.

> We had our big shower at Number Ten South Wellington. You come out in your dirty clothes and you go and have a shower and someone would wash my back and I washed his back and the whole place was full of naked men and no one paid any attention.

Although smaller mines in Alberta had had them for years, washhouses were not mandatory in B.C. until after 1940. That made Granby's washhouse all the more unusual. No one knew for sure why the company was so generous, although a skeptic opined that it was all a tax dodge. And the luxurious facilities at Granby were small compensation for the danger of working the Granby mine.

Within three years of its opening, some people were calling it the Slaughterhouse. One hundred blowouts, that lethal combination of coal and methane gas under pressure that can dislodge hundreds of tonnes of coal at once, occurred in 1921 and 1922.

> Now a blowout is different from an explosion. A blowout is gas forcing the coal out, but it's not ignited. It's in behind and it bursts the coal out. There really isn't a warning before a blowout, but you can tell. When you get your pick in as far as the handle will reach, you can feel it if it starts to bump. Bump. Bump. It's time to back off.

> It always gave a warning like a tapping up on the roof. My brother and I had blowouts, two or three of 'em, but we never got caught off guard. We always beat it when we heard this tapping in the roof. And it would blow out maybe thirteen or fourteen sets of timber. You wouldn't believe it.

A blowout could bury a man and crush the life out of him or the released gas could asphyxiate him. Sometimes the coal blown out was as fine as powder and could suffocate a man buried in it. But the old hands liked blowouts. They could detect one ahead of time, get out of its way, and then, after the gas cleared, load the tonnes of coal it released. A man could work for days filling cars as fast as he could shovel – no blasting necessary. For men paid by the number of cars they filled, a blowout was a bonanza.

Years after Granby mine closed, a miner in Number Ten South Wellington found himself working very near the area of the Granby blowouts. Mine inspectors experimenting with ways to handle potential blowouts had ordered the fire boss and his crew to drill two shots in a "V."

> And the fire boss brings an extra cable and we fired them two shots. He says to me, "Now just stand here and if anything happens run down and warn the other men. We stood there about half a minute and woof. My partner says, "I hope it blows enough so we don't have to shoot any more." Boof. It went again. Boy oh boy. I thought the world had come to an end. It knocked out fourteen sets o' timber, knocked down all the brattish boards and it lifted me off my feet and the roof cracked above my head. And we run. We ran down and warned the other men and the gas kept crawlin' and they had to board off the main slope and chalk it up "Danger. Keep Out." And we went back there two days later and there was about seven hundred ton that blew out but the roof held.

Dangerous or not, the coal was of good quality and the miners liked to dig it. But there were special rules to follow in a mine prone to blowouts. Explosives could not be used, so all the digging had to be done by hand. Miners had to drill three holes six metres ahead of the face that would, in theory at least, drain away the gas. Pull bells at frequent intervals warned the men to proceed to the nearest fresh air in the event of the release of methane gas; generous amounts of non-explosive dust covered all travelling ways. By 1925, the new measures had reduced the number and intensity of blowouts to twenty-eight for the year, but the inspector noted that "a great deal depends on the miners themselves, as they are in most cases very careful, but some of them do take chances owing to familiarity with the danger."

And Granby mine still claimed lives. The triple blast of the mine whistle that announced a fatal accident sounded often across the hills. The steep 760-metre slope carried death on its double set of tracks as runaway cars caught miners off guard. On the twin slope or "manway" running parallel to the main slope, a stretcher bearing a dead man to the surface was not an unusual sight. The men called them the killer slopes.

> My brother and I were working at Granby and he told me, "I'm kind of scared of the job I'm on." He was taking the rope after the trip passed. The trip would have six cars of coal and then two timber trucks on it and he says, "I don't like them timber trucks." The coal cars had a drag on them that would throw the cars off the track if the rope broke, but the timber trucks didn't have a drag and they weighed about fifteen hundred pounds and the slope was steep. So anyway I says, "If you're scared of your job, you quit. I'll keep ye." "Well," he says, "I'll quit as soon as I get my suit paid off." Well, by golly, the very thing he was scared of happened. His last timber truck the rope broke and the truck come down and the undertaker told me there weren't a whole bone in his body. It carried him about a hundred feet down the slope. When they picked him up his light was still burnin', but he was dead. He was only twenty.

> It was a dangerous mine but it was a good mine. Too much gas. That was the only drawback. That's why it's left today. They spent a lot of money going through a fault way down and after they got through and they were back in the coal again they discovered that the seam they got into possessed more gas than the one they were working. It was really too dangerous and I have an idea the government encouraged them to put a lid on it.

The Granby mine was only seven years old when extraction of pillars began. The opening of a second mine on the property in 1929 did not improve the output. The company had spent an enormous amount of money on a mine that lasted only fourteen years, In September 1932, Granby closed. The houses were sold and taken to Ladysmith and Nanaimo. The doctor's ten-room house now sits on Vancouver Avenue among gracious homes from Nanaimo's past. The sports field disappeared in truckloads of gravel for a highway project. A few foundations remain and a huge Douglas fir grows in the middle of the killer slopes.

Many people remember the houses in Granby because they were so much finer than other miners' homes. The houses in Extension and South

Wellington, or on the tar-flats of South End in Nanaimo, or even the company houses of Bevan and Cumberland did not have the amenities that the Granby houses did. Few had indoor plumbing or hot running water. Houses in the mine camps had electricity only if the mine did. When the mine closed, the miners lost their jobs and their homes lost their power. But all miners' homes, whether in Granby or Bevan, had a garden. Not always a lawn or flower beds, but always a vegetable patch. Vegetable patches fed large families on small incomes.

With the exception of those who lived in company houses in the Cumberland area, most twentieth-century married miners owned their own homes. It was a matter of independence. A miner who owned his own home was free to work in the bush if there was no mine work available. A significant number if Nanaimo area miners settled the Five Acres district, which allowed them to combine mining and farming and achieve a larger measure of security.

Vancouver Island had always attracted married miners. Since the earliest days of the HBC, company recruiters preferred married men. They knew that once a man had moved his family to the Island it would be difficult for him to leave if he became dissatisfied. Island coal camp families were generally stable with an absence of such social problems as child neglect and wife abuse common in coal camps on the prairies. The men who arrived single sent home for a woman to join them or set out to find a wife on the Island.

I was going to a miners' meeting in Ladysmith and this girl was at the side of the road with a flat tire. I knew who she was and she knew who I was, but we hadn't met. So I pulled up behind her and said, "Trouble, eh?" "Flat tire." I had a pair of overalls. I pulled them on and got my tools out, jacked up the wheel, patched it up and took the spare out and fixed that one. It took about twenty minutes. Then she said, "There's a dance in Granby Saturday night." I said, "I don't dance." So she says, "There's a good show in Nanaimo Saturday night. How about taking in a show?" That was her last dance.

We both lived in South Wellington all our lives and I guess he got tired of running around with every Tom, Dick and Harry and I was just a young innocent girl and I guess he picked on me. Oh, he was quite a high stepper, don't worry. I used to see him all the time, but I never thought I was going to end up with him. Next year we'll be married fifty years.

I didn't get married until I was forty-five. I didn't wanta get married before that and I didn't really wanta get married when I did. But she claims she couldn't have no children so I thought well, just the two of us, it'd be all right. By God, you get the darn thing started you couldn't shut it off. We had six children. We got along. I don't know if it's the Lord or who it is but somebody seems to help you out.

People helped each other and people helped themselves. Anyone short of food or fuel would usually find a bag of potatoes or a bucket of coal on the porch in the morning. When a baby was due, the neighbour women officiated, and when a mother was sick, a relative or a friend fed and cared for the children.

> There was an acceptance of characters. In our neighbourhood there was a man who was not too fond of work. He had a big family so everybody helped him out. When he'd go hunting he got lost every year. He used to just sit until somebody went and found him. There were other characters who weren't too fond of working. There was one man, and his wife would look at him asleep in the morning and she would say, "Oh he's too peaceful to wake up and go to work." He couldn't get a job anywhere. He died just a short time ago at about ninety years of age. None of us had that much, but we shared. We were poor but we didn't know it.

> At first the miners used to get a ton of coal a month for nothing. That was given to them. And then in the summer time they didn't use so much so for a few beers they would give the beer parlour their coal order. Well the company got to find this out and then of course started to charge them three and a half for a ton of coal. Well, you couldn't blame the company. And it was the same when it come to Christmas time. They used to give turkeys away. A man would go into the company office and he would squawk, "Oh, he got a bigger turkey than me and I got more kids than he's got." "Well," the company said, "we'll put a stop to that." So they stopped the whole thing.

Family life in the coal camps had survived the war and the strike. When the war ended in 1918 and the men came home, coal production was high and there were three new mines to absorb the extra manpower. There was a determination to put the war and its bitter memories behind and get on with living. But that fall an influenza epidemic spread around the entire world. British Columbia was especially hard hit, with over 30 percent of its population ill and twenty-three people per thousand population per year dead in Vancouver. It may have been that the virus came home with the returning soldiers, but no one could prove that. Nor could anyone explain why the disease was particularly hard on people aged twenty to forty and on pregnant women and coal miners.

> The only thing I haven't forgotten and I never will was the flu we had in 1918. I was just a young kid and saw 'em throwing those dead people into those buildings. I'm telling you I never have gotten over it and I've been scared of dead people ever since. They turned black and died.

> When the flu epidemic hit Cumberland I got it and they had to open up the schools for hospitals. I was up there in the bed and they were dying around me like flies.

> During the big flu in 1918, I was working in Harewood mine and you would go to work and some men would make it into the mine, but we would have

> to pack them out and that went on for about a week. They were fine when they went to work and they just got sick like that. One fella died in the mine. He died before we got him out. In this neighbourhood there was a death in every third house I guess. An awful lot of people died. They didn't give it a name at that time but they died just the same.

In the spring of 1919, the flu epidemic ended just as suddenly as it began. But another legacy of the war remained. This one made no one sick. Miners still dug coal and the sun still came up in the morning. But something had gone out of people's lives and that something was alcohol. Canada was in the grip of its brief flirtation with Prohibition.

Yielding to pressure from prairie temperance groups, the federal government had used the War Measures Act to prohibit the sale of alcohol for one year in 1918. In 1919, politicians made Prohibition the law with the Canadian Temperance Act. Although bars and hotels were dry only until 1922 in B.C., many of the coal miners of Vancouver Island remember the four years of Prohibition as if they were an eternity.

The only legally available alcoholic beverage was near beer, which contained 3 percent alcohol. A drinking man might try to persuade his doctor that he needed an prescription for alcohol to "cure his cold," and failing that, he had to make his own booze or seek out the local blind pig or bootlegger in keeping with time-honoured Vancouver Island tradition.

> A bunch of us got together one time and bought a bottle of the stuff and I took one drink and said, "That's enough for me, you can have the rest." A lot of people went blind drinking that moonshine. Wasn't properly distilled. Prohibition didn't last long here. There was a lot of that going on in the States and they came up here to get it.

Alcohol was not a necessity of life, but food was. Most people grew their own vegetables; those with farmland produced milk and eggs for their own use and for sale; and most families had a hunter and a fisherman. Small stores filled the balance of grocery needs, making weekly deliveries in horse-drawn wagons and allowing their customers to buy on credit. Bills were settled on payday.

Saturday was the day everyone came downtown. And every second Saturday was payday. Miners would meet their working partners at the Royal Bank, their shared pay stub in hand. The company accountant would stand by the teller's wicket, examine each pay stub, and identify the miners to the bank teller who would then issue an envelope of money for each stub received. Bending over the bills and loose change, the partners would divide the money between them, and before leaving the bank each would pay his fifty-cent or $1 premium to the Sick and Accident Fund or the Hospital Treatment Fund. There might also be someone collecting donations to help a new widow through a rough time, then it was out onto the street where children clustered around their fathers knowing that some of the loose change might come their way. The women took money and went to pay the merchants who had given them credit and the men went to the barbershop. Then, shorn and respectable once more, the men who drank headed for the bar.

It was a comfortable Saturday ritual bought by hard labour in dangerous conditions. Demand for coal and labour might vary, the cost of living might fluctuate, but in the years between 1900 and 1937, the amount of pay in those envelopes stayed virtually the same.

> The wages were very poor. When I started in the mine in 1915, a kid on a winch, I got a dollar and a quarter a day. And a driver got $2.86. The miner's wage if he wasn't on contract was $3.30. Course if a miner was on contract he could maybe make more but they made sure, see, They kept the price low. When I started in the mine, a contract miner got 91 3⁄4 cents a ton. Per ton. When the mines finished we were loading by the car a dollar a ton. The drivers were getting $6 at the end but that was all. They weren't overpaid.

And the company had already deducted money from the pay envelope for the supplies the miners had bought in order to do their jobs. A four-pound can of blasting powder – $1; each cap used by the fire boss to loosen the coal – ten cents; lamp rental for two weeks – twenty-five to thirty-five cents; a tool bought at the company store – another deduction.

Miners could make more money on contract if they were smart and worked hard. Working in teams, these men could make a comparatively large amount of money if the roof was not too low, if the drivers kept them supplied with enough empty cars, if the seam did not pinch to nothing, if the fire boss was there when they needed their shots fired; if the roof held. Some men preferred to work for a basic rate of perhaps $5.25 per day rather than risk all those "ifs," but the drawback to being on basic rate was that once a man had filled enough cars to add up to $5.25, there was no incentive to dig any more coal. Why give the company free labour?

Contract work could be more dangerous, however. Because the contract miner wanted to get out as much coal as possible, it was tempting for him to ignore certain safety measures such as timbering, for which he was not paid. The time taken to timber was time away from loading coal for which he *was* paid. After the union was finally recognized, a contract miner was paid for every timber he installed, for time spent brushing to make the floor lower to accommodate cars, and for company work. Ninety-five cents a ton for a contract miner or $1.40 per day for a boy on the pithead – with these wages a man got by, but not magnificently.

> I worked Timberland mine at McKay Lake and I travelled from Northfield, thirty-seven miles return a day on gravel road by bicycle. I done that for two weeks for $2 something a day. I was running hoist. Well, $2 a day then you

could buy a can of snuff for fifteen cents, eh? You know what it costs you now? Pretty near eighty cents. Your prices were low. We paid $7 a month for a company house, which included water and you got your coal pretty cheap. On the haulage if you got $90 a month you had big pay. In those days your wife didn't work either. Just you and the kid when he was old enough to go down the mine.

They had a commission here looking into how the miners managed to get by and they asked the head surveyor, "How can a man get by?" He says, "The miners are by nature very frugal." Well yes, I'd like to have been there. I'd have told him we had to be frugal. I never saw a rich miner yet.

There was a day in 1924 when several out-of-towners decided to become rich at the expense of the miners. It was December 12th, the day before payday, and everyone knew the Royal Bank had its safe full of money for the Western Fuel payroll. At 2:30 p.m., a seven-passenger Durant pulled up in front of the door. While the driver waited in the car, six men got out, one well dressed and with a confident air, the others wearing khaki pants and mackinaws. In a moment they were in the bank, guns trained on the tellers. Two of them guarded the door, two made the staff lie on the floor and the other two stuffed all the tellers' paper money – $30,000 to $40,000 – into a sack. Then they went for the money in the safe. Knowing the manager would have the combination, they burst into his office only to find that he was out. The manager had a dentist's appointment.

The bank robbers decided they could wait. New customers arriving at the bank got a curt wave of a gun and an order to join the line-up of hostages. Twenty-five minutes passed. Finally the leader could wait no longer. The gang ran from the bank, jumped into the Durant, and left the much slower police car in the dust and $88,000 cash in the safe.

The thieves abandoned the Durant at Boat Harbour and jumped aboard the *El Toro*, a fifteen-metre speedster launch that was waiting to whisk them away. All the rest of the day and through the night, the Canadian navy, police boats from Vancouver and Victoria and private speedboats and seaplanes searched for the *El Toro*. But it was eleven days before Canadian and American police caught the robbers in Seattle and the identity of their suave leader became known: he was a former Seattle police detective.

Thanks to the bank manager's dentist, the miners received their pay on December 13th and the ritual haircuts and visits to the beer parlour went off as usual.

The beer parlour. Naw. It was called the bar them days. Yep. You walked up to the bar and put your foot on the rail and your arm up on the bar and there you were.

In the 1920s when coal was still king, Nanaimo had 10,000 people and twenty-eight bars and beer parlours. Take away the children, the non-drinkers, and the women, who seldom drank in public, and that makes approximately one bar for every eighty men. With only eighty men to keep each bar in business, competition kept the price of beer to five cents a glass including all the sandwiches a man could eat. Even when the price went up to ten cents a glass in the 1930s, the big sign on top of the Crescent Hotel still read Five Cent Beer.

There was a theory that beer drinking was beneficial, that beer would wash the coal dust out of a man's throat. Some bars would give a miner still dressed in his work clothes a free first drink. He had to get the coal dust out.

> You don't know how much they just used that for an excuse. I can't speak highly enough of the type of person that the coal miner was. There were a great deal of Scotchmen and Welshmen here and they were very fine, thrifty people. They were honest as the day is long and the only trouble was that they perhaps liked their beer a little bit too much.

> When I was a little chap I can see them now. Coming home so drunk they couldn't hang on to the fence. Matter of fact, our neighbour across the street was literally so intoxicated he couldn't hang on to the pickets of the fence. It was tragic, because when he was sober he was a good father, a good husband, and a good provider. But this did occur periodically and that's where the money would go. I think drinking to that intensity has gone out.

Despite the tendency of some to overdo it, most miners were more temperate – if still devoted – beer drinkers. The bars of South End were particularly popular because of their proximity to Number One mine. The Balmoral, the Patricia, and the Columbus were ideally situated to save their clientele too arduous a walk in their thirsty condition.

> When I was driving in Number One, one of the diggers would give me two bits just before we walked out with the mules. So we'd get out of the mine before the diggers and we just popped over the hill to the bar and got a two bits bucket of beer. By the time we got back, the cage was up and the miners would get the beer.

The two bits bucket of beer was popular with those who preferred to drink at home too. They dropped a quarter in the dinner pail they took to work and sent one of their children down to the nearest bar with it. The child passed the bucket through the door to someone inside, who took the quarter, filled the pail with beer and handed it back through the door to the waiting child who rushed it home to where his father and perhaps several of his buddies waited to dip their glasses into the suds. To some miners, two things were important after work, a beer and a bath, but the beer came first.

Downtown Nanaimo was where there were the most bars: the Terminal, which used to be the Eagles; the Lotus, which used to be the Temperance; the Queens, which used to be the International; the Globe; the Occidental; the Commercial and the Old Flag. Farther away from the centre of town across from the arena, the Newcastle would serve a sports fan a quick beer at half time. The next stop on the road to Wellington, where Robert Dunsmuir had first mined coal, was the Quarterway, better know to some as the Talbot. The Halfway in Northfield was halfway to Wellington and the Wellington Hotel was all the way – the only bar left of the ten that had done business there before the Wellington mines closed in 1900.

Every mine camp had at least one hotel within easy walking distance of the pithead. The Tunnel Hotel was the only one left in Extension after James Dunsmuir ordered the miners to move to Ladysmith, but it served its patrons well. It was homely but its polished wooden dance floor was second to none. The pub in South Wellington had survived the fire in 1914 while everything around it burned to the ground, a fact that many serious beer drinkers saw as some kind of divine intervention. The bars that lined Dunsmuir Avenue in Cumberland were defiant when police tried to trap them in a violation of the Sunday drinking laws – the Cumberland bars stayed open seven days a week.

Because many of Ladysmith's inhabitants were family men and the mines they worked in were some distance away in Extension, the town's bars were more circumspect. But the men were also hard-working miners and some of their colleagues were bachelors and many of them liked to drink beer. In 1915, there were fifteen hotels providing room and board to bachelors and selling beer.

> There were fifteen hotels and every one of 'em was doin' business. And there was also a brewery and another place up on the hill that was a wholesale liquor place. Beer was $1.50 for a dozen big quarts. When I was fifteen I worked at Simon Leiser's General Store and he sold a bottle of Scotch for $1.25. Beer was all five cents a glass. One of the men I worked with sent me for a bottle of whisky for him every working day of the week. Of course, it suited me to go. I'd sneak a smoke and then rap on the door of the old Frank Hotel and they'd bring me out a shandy while I was waitin' for the whisky.

For the men working in the black dust and dim light of the mines with the constant spectre of an explosion or a cave-in with them day after day, it is not surprising that they needed a few beers to banish unpleasant thoughts, for a while at least. The Bayview, the Portland, the Columbus, the Pretoria, the Victoria, the Traveller's, the Cecil, and the Abbotsford, the King, the Palace, and the Pilot, the Europe, and the new Western all provided a little respite. The Temperance was there for the men who didn't drink alcohol.

The Extension Hotel, one of the ones James Dunsmuir had ordered moved to Ladysmith from Wellington with a stop in Extension, was a fancy place in its heyday. A heavy oak bar ran half the length of the building with a big mirror on the wall behind that reflected the back of the beer taps and liquor bottles on display. There was a poolroom next door and a table for cards, but in the main room the serious drinkers stood at the bar with one foot planted on the rail. On the floor between the rail and the bar ran a trough of running water convenient for spitting. At the top of the hill the Queens had a trough too and a bartender who could shoot a glass along the bar just as if it were a shuffleboard. The troughs disappeared later, probably on the orders of some outraged public health inspector.

There were fights now and then in the coal miners' bars. Fighting in the mine brought a $50 fine, so scores had to be settled outside. And there was always the occasional man whose drinking got in the way of supporting his family. It had always been so among coal miners and the measures taken to prevent a miner from working while drunk were recognition of this. Fire bosses checked for alcohol-scented breath at the beginning of shifts; drink was strictly forbidden in the mine. There were temperance societies in the camps with Finns at the forefront in these self-help organizations. But for the average miner, the bar was his club and Saturday was his day.

> We always went to town on Saturday. You stayed until you seen a show in the evening and then you'd come home. Oh there'd be a big bunch of us altogether. Them days you did a lot of walkin'. Like the whole of Departure Bay would walk in. In the winter there'd be about eight or ten families there, maybe a dozen. Not many. Just enough to keep the school goin'. So we'd all go to the show and it'd be dark when we came out about eleven. Then we'd stop in at the bar and have a bracer to come home on.

> Saturday was the day we all got paid. Then we'd go off to the bar and I think there was more coal mined in that beer parlour than was dug in the mine. It was the only thing you know. You worked in the mine and you were a miner. We'd start a conversation up and maybe someone else would come along and join in and they were just haulin' coal and arguin'.

The bars all had closing time, but there was a place where the after hours crowd could go to drink, play cards, and visit a prostitute if they had the inclination. Fraser Street, a steep road that ran along the tidal ravine, was the address of five or six houses that made up the red light district of Nanaimo. Madams and prostitutes plied their trade legally under strict rules set by city council. The bylaws required each prostitute to visit a doctor every Monday and on the days when she went to town to shop, to be home by 5:00 p.m.

The reputation of the street provided much topic for conversation. Most citizens agreed that it probably made the streets safe for respectable women. Young boys bragged to each other about having to deliver groceries to the houses; little girls sped guiltily through on their bicycles; and mothers pretended it didn't exist.

> Well I don't remember the grown-up attitude to Fraser Street, but I remember the teenage attitude. Some of us girls when we were at school used to go to town once in a while and we would think we were very brave if we ran down Fraser Street as hard as we could go. It was an evil place.

> I didn't know what they meant when they were talking about the red light district. I do remember going there on my bike with a friend and one day we felt especially bold and we came down on Fraser and some woman yelled, "Have you seen the ice man?" and I got so scared I nearly fell off my bike. Mother said, "Don't go down there again. Little girls don't ever go down in those places." I didn't know what it was about.

> Before I went in the mines when I was a kid, I peddled groceries on a bicycle. Pretty near every store had a bicycle in those days and I used to have to go down to Fraser Street with the groceries. There was one woman there and I'll never forget her saying to me, "You're getting to be a big boy now and I guess you'll be comin' in to see me one of these days."

The clientele of Fraser Street was not nearly as willing to admit going there as were the children who went through on a dare. Some will admit to an after hours drink or a game of cards but few will admit to any other activity.

> The only time I was on Fraser Street was when I was in my teens and I was a messenger for the express office.

> Miners went there mostly to drink; it was the loggers that went there for the girls.

> It was mostly the sailors off the boats, not the local people.

> Just for big shots, business guys used to go there more than anybody else.

> If it wasn't for the mine men it would have had to close a long time ago.

It finally had to close during the Second World War. The huge number of soldiers brought in for six weeks of training at the army camp turned Fraser Street into a law enforcement nightmare. City council ordered the houses closed and the women dispersed. The talk afterward evoked the image of the "hooker with the heart of gold." People remembered how respectable and law-abiding they were, how generous in their contributions to charity especially during their busy Christmas season and how the women of Fraser Street paid for the furnishings in two rooms of the old hospital.

A philanthropic institution of a much different sort had established itself right in the midst of the area where most of the miners lived. The church on Haliburton Street served three denominations: Anglicans in the morning, Presbyterians in the afternoon, and Methodists in the evening. The Methodists in particular reached out to the coal miners and ministered to their social as well as their spiritual needs. During the strike it was the Methodist ministers who gave unqualified support to the strikers. Reverend J.W. Hedley wrote one of the few accounts of the first year of the strike from the striker's point of view. When he was transferred away from Nanaimo shortly afterward, however, his parishioners wondered if the Methodist church hierarchy shared his views.

Music filled the coal miners' church. The Welshmen of the parish were the mainstay of choirs and bands, but music was important to all its parishioners. Every year a conductor imported from Vancouver directed a performance of *The Messiah.*

The church was the focus of some people's lives. There were young women who attended Sunday school until they were twenty-one years old. Such devotion was not felt by the majority of young boys, however, many of whom scrambled joyously over the hill to the Cricket Grounds to watch baseball and soccer matches after Sunday school, their Bibles and their Sunday clothes marking them when the Sunday school superintendent took his weekly post-devotional walk in the same direction. But however reluctantly, more children attended church than parents.

> My people were like most people. They went to church now and again. But when we were kids, they made us go.

> They didn't go much on religion, but I think most of the miners were God-fearing. They worked hard and they drank beer and they swore and they did things any miner would do, but they were basically a good, hardworking, clean-minded group of people.

> I was supposed to go to church when I was young, but when I got old enough I said, "To hell with it." I got fed up with it. I believed in enjoying myself. Mind you, nobody will say anything against me because if I could help someone I always helped 'em. That's my religion.

Nanaimo was not particularly unusual in its contradictory attitude towards religion. Every town has its devout people and those who can't be bothered – often within the same family. In addition to the Anglicans, Methodists, and Presbyterians, the latter two of whom joined to become the United Church in 1925, there were Roman Catholics and several smaller religious groups including Plymouth Brethren, Salvation Army, and the Pentecostal Assembly.

> We'd speak on street corners. So did the Salvation Army and the Plymouth Brethren. We'd have outdoor meetings. Our corner was by Spencer's store downtown. There was the odd occasion when a few rocks and the odd tomato came down. It led to traffic problems when the automobiles came in, but in the horse and buggy days it wasn't a problem.

> My mother and father were religious. My mother would drive you crazy by religion. You come in at night after doin' a hard day's work and you got down on yer knees and you had a prayer and then they'd read a chapter out of the Bible. When I grew up I said, "Okay, that's all I want to hear now."

The inhabitants of the old coal camp at Wellington had reason to be skeptical about religion and its proponents. The story goes that a man who said he was a minister turned out to be a taxi driver from New York who had stolen a minister's credentials. By the time he had been in Wellington for a year, a number of unexplained pregnancies had blossomed. The miners had tarred and feathered him, run him out of town and told their wives not to talk to any minister, which made it difficult for the taxi driver's replacement, a bona fide member of the clergy, the Reverend Campbell-Brown.

> When Mother and Father arrived in Wellington they tried to visit the miners' homes, but no one even answered the door. In the course of a year or two they broke down the prejudices of the area and became very much a part of the community. When my father died in 1924, the miners took over the funeral arrangements. They had a community working bee to clean up the neglected old burial ground and put up an elaborate marble and granite monument.

A popular minister, a fervent belief in God, a love of music – all these reasons and more brought miners and their families to church. But the love of music found expression outside the church as well. The people of Courtenay, Comox, and Cumberland joined to form a large mixed choir. Nanaimo had a Welsh male chorus that appeared in festivals and competitions all over the Pacific Northwest. When one Welsh family came to Vancouver Island in 1912, one brother mined coal to support his family and played in the local band for fun; his brother was the bandmaster but supplemented his income working as a miner.

Every community of more than a few hundred people had its own band. The Silver Cornet Band – begun in 1872 as the Nanaimo Brass Band and known in modern times as the Nanaimo Concert Band – made its mark on the memories of Island miners. Whether it was marching in a May Day parade or a mass funeral procession, the band played a prominent part in the ceremonial life of the community. The measured beat of muffled drums conducted thirty Extension miners to their graves in Ladysmith and sixteen Protection miners to the cemetery in Nanaimo, as the band played the "Dead March" from Handel's Saul. Having escorted their friends to the graveyard, the musicians would remove the black muffles from the drums and strike up the rousing and sometimes ribald ditty, "The Girl I Left Behind Me" reaffirming in their incongruous choice of music that in times of great sorrow life must carry on.

In a time before radios and phonographs brought music to every home, the music hungry crowded around the bandstand on Front Street for Band Sunday when two or three bands took turns presenting their repertoires to an appreciative audience. Mining families did not confine themselves to merely listening to music, however. Coal towns from Ladysmith to Cumberland were well-supplied with dance halls. To the music of combos like the "Novelty Five" or "Bennie's," the young and the not-so-young could cavort every Saturday night to the combined sounds of a piano, a violin, and a few brass instruments, played by men who in real life were miners and merchants and doctors. To "hard and fast music very much like a jitterbug nowadays," work-weary men and women shed their troubles. Not every man liked to dance, but there were some truly devoted exceptions.

> Oh, I was fond o' dancin'. And you know what made me a dancer? I used to work in Brechin mine when I first started and the two fellas that I had down there workin' with me, they were dance teachers. And they learned me to dance and I never quit. We danced in the Assembly Hall in Nanaimo where the Athletic Club was. Different fellas played the music. Accordions mostly. Used to go to Nanoose Bay fer a dance, oh yes. Then horse and buggies come out. Oh, we used to go all over dancin'.

There were dance halls at Northfield and Wellington, Harewood and Ladysmith. Richard's Hall in South Wellington used to sway when the dancers moved in unison. The Finn Hall in Chase River and the Tunnel Hotel in Extension were local favourites. In Nanaimo you could dance in the basement of the Oddfellows Hall or at the Pygmy with its well-waxed wooden floor sprung so it bounced gently in time with the dancers.

And there was dancing at Happyland on Protection Island in the old pavilion where the Japanese fishermen used to dry their nets and store their dories. Or dancing on Newcastle Island at the annual Miners' Picnic. The dancing was the finale of a day of fun that began when the coal company tug, the *We Two*, pushed scows loaded with picnickers to the wooded island in Nanaimo's harbour. Each family brought enough food for the whole day, packed in a large clothes basket or some other commodious container. Men pitched horseshoes and played softball and football; kids played catch; women competed in "thread-a-needle" and the egg and spoon races or just spread the picnic blankets and talked.

> And you could fire a shot in Nanaimo and never hit a soul. The men used to pay so much for this picnic and the company paid so much. The kids got ice cream and all the pop they could drink. The men got all the beer they could drink and then there was a big dance at nighttime on a big flat deck out in the open. You'd have a live orchestra. The *We Two* used to go over and pick up the stragglers that missed the last scow.

The best beach in the area was on Northumberland Channel just before Dodds Narrows, but only people with a boat could get there. William John "Skipper" Moore had a boat that would carry forty people and there were that many people from his British hometown of Whitehaven, Cumbria, living in Nanaimo's South End. Wilsons, Littles, Quails, Swinburnes, Bamfords, Rotherys, Branches, Twentymans – they were all Whitehaven people. Skipper Moore would bring his boat from its moorage in the Nanaimo River Estuary to the landing near the coal wharves on Cameron Island, and he'd whittle while he waited for his passengers. Soon they'd appear, walking along the coal company train tracks in single file, everyone loaded down, their tatty pots wrapped in towels to keep the food hot. When Skipper Moore died in 1944 from the after effects of an accident in South Wellington's Number Five mine, his family and friends remembered those excursions and how the steady thump of the boat's Easthope engine propelled them through the water, and how fifteen children could fit on the cabin top and how they sang all the way to the picnic.

Cumberland mining families went to their annual picnic at Union Bay on board the Canadian Collieries train, which stopped for passengers at Bevan and Cumberland, the holiday dress of the passengers transforming the homely railway cars. Gone were the usual drab work clothes and housedresses. Instead the men wore white shirts and their good hats, the ladies their best summer dresses. Children started out the day with scrubbed faces and clean clothes.

There was a picnic on every summer holiday but the biggest holiday of all was on May 1st. May Day was the original Labour Day, more acceptable to working men than the September holiday created by government edict. The holiday had fallen into disfavour after the Bolshevik Revolution in 1917 when governments and police forces were watching the working class warily, convinced that the communist revolution would spread all over the world. To fill the void left by May Day's fall from grace, the predominantly British population of Vancouver Island embraced Empire Day. Parades and picnics, sporting events, and dances on the twenty-fourth of May weekend celebrated the birthday of Queen Victoria and her glorious empire, even then in the process of coming apart.

But picnics weren't the only thing. The miners remember whist drives and silent movies with Charlie Chaplin, Harold Lloyd, Xavier Buckel, and Pearl White. They remember the first talkie in 1929 and photo night in Cumberland. They remember themselves as kids on Saturday afternoon when they came to town for fish and chips and a Tom Mix movie. Every business in town had an ad on the screen before the show began, then the record player started the music and the black-hatted robbers flashed on the screen bent on robbing yet another train, each child in the audience knowing that the white-hatted good guys would always win in the end.

They cheered the local guys on as well. Soccer football was serious business. Teams from Cumberland, Granby, South Wellington, Nanaimo, and Ladysmith played off in spring leagues, summer leagues, and winter leagues, many of the teams going on to greater sports glory. Nanaimo United won the Dominion championship three years in a row; the Junior Southend Foresters were three-time B.C. champs; the Ladysmith team had good years from 1908 to 1910 and from 1921 to 1925; and the Cumberland players were champions too.

Before 1924, the Nanaimo teams played on the Cricket Grounds and after that year, on the Central Sports Ground. For a fee, spectators could sit in a covered stand near Scotchman's Bluff, so-called because people of a certain nationality had a reputation for avoiding the entrance fee by sneaking into the ground by way of paths that skirted the rocky outcrop. Normally the stands held sixteen hundred people, but at the final game between the Southend Foresters and the Vancouver Excelsiors over two thousand people crammed in to watch.

Most of the players were amateurs who made their living digging coal, but some were coal miners by convenience only.

> If you played football or were in the band you got a job right off the bat. Yeah, and an easy job too. They were swell on the guy and give him a job off the top and they'd go swimming for the afternoon. A job doin' nothin'. Played soccer.

> Most football players were Scotch and in the mines where the bosses were Scotch they were especially avid for football. The miners would come from Scotland and they were guaranteed a job if they were good soccer players.

> If you were a good soccer player they found you a job. And you got reasonably well paid and they used to pay them so much, $10 if they won a game, $5 if they drew a game and nothing if they lost.

> They always talked football, football and every weekend got them all hot and bothered to go to these football games and there was nothing else, that's all

we talked about. It kept us happy looking forward to football games every weekend.

Football and baseball were the big things in Cumberland. The managers figured if they kept the people interested in something else besides work, they'd keep them happy. It was the same in Nanaimo. They used to come up here in special trains to play us.

Baseball players were treated similarly. Employers lured semi-professionals to the Island with jobs in the woods or the mines. But their primary job was to play ball. Ball players in Granby went to work with their ball gloves on their belts. A baseball player received only basic pay, but if he missed a day he still got paid. One shortstop knew he could call the manager if he needed extra money. When his playing days were over, he quit the mines.

Despite the favouritism shown the athletes, some of whom knew nothing about mining, and despite the knowledge that good rousing football games were a management ploy to keep the men quiet, there was no apparent resentment of the players by the real miners. The games provided such release for player and spectator alike, their exuberant physicality such a contrast to the restrictions of the mine, that all viewed the system with tolerance. This is remarkable considering the apprenticeship the regular miner had put in so he could work at the face.

You had to work there so long and then you got what they called the miner's ticket. I think they cost $1 them days. You had to work so long down the mine, learn all of it and then you'd take the oral exam, get your ticket and you are a full-fledged miner then, I guess.

They'd ask you all these different questions and what you would do if you walked into a place and you found gas, and how much powder are you supposed to carry with you and are you supposed to fire them shots? No. It's up to the fire boss. But you can tamp your powder and stuff the powder in your hole and then you got a little sandbag, you use a copper bar to tamp that in. You got to be so doggone careful. You don't want to set off an explosion. They were electric blasting caps and anything is liable to set them off. I used to go in the bush and cut myself a stick and bring that down with me. I figured that was the safest thing to do. I wasn't going to trust copper anyway.

My dad used to say your basic three elements of school are math, science, and geology and they all play a part in mining – coal mining especially. You gotta be able to read the ground, maybe be a bit of a scientist to figure out the structure of it and geology. And mathematics. You gotta be able to calculate the risk in the areas where you gotta go in to. Coal mining you got the wall squeezing on your other side, the roof's coming down on top of you – you gotta brace.

A miner himself has got to be very skilled and just because he doesn't have an education in school doesn't mean he wasn't skilled labour. For instance,

> my grandfather never went to school after he was seven years old, but there never was a skillder miner anywhere around than him.

Experience was what counted most. By the time a boy had worked at the various jobs starting with the picking tables and then going below to be a winch boy or rope rider or driver, he had acquired a measure of knowledge. Good drivers sometimes found their progress to the next level stalled because they were so valued for their ability to handle animals. The best job was the digger's because he received the highest pay, especially if he was on contract. And the best way for a young man to learn the digger's job was to work as a backhand loading coal for an experienced digger – usually his father or older brother.

A man with a desire to upgrade his work situation and increase his responsibilities could become a fire boss if he could qualify for a third-class ticket. This involved taking a correspondence course or going to night school and passing a five-hour written exam. A fire boss had to hold a St. John's Ambulance First Aid certificate and a Mine Rescue Competency certificate both of which required taking additional courses. Having received a third-class ticket, a man could oversee ten or more men and fire all their shots. He was responsible for their safety and for administering first aid.

> I passed the exam and got my certificate, but there was a fella ahead of me. The mine foreman came to me and said, "We want you to be fire boss." I said, "Well, what about the other fellow?" "Oh," he says, "he's no good." I said, "After I was fire boss you'd say the same thing about me. I don't want the job." So I don't know how long the other guy was on the job maybe six months to a year, but he got caught by an inspector. You're only supposed to tamp one hole at a time and the inspector caught him with four holes tamped. So it cost him $50 and they cancelled his ticket. Then they come after me again and said, "Well, you'll have to go fire boss." And I said, "Okay. I'll take the job now."

A good fire boss was a man in the middle. He could gain the respect of his men by quick action in an emergency and by fair distribution of the shots he fired. He could gain the respect of the company by ensuring maximum production and minimum time lost. Standing by the cage at the pithead, checking the in-going shift for pipes or liquor-laden breath, searching for a forgotten match head in a pocket corner or a forgotten cigarette behind an ear, checking the lock on a battery box or an older safety lamp, he was isolated from both the men and the management. This isolation was more apparent after the company recognized the union: management classified fire bosses as daily wage workers and not eligible for monthly salaries, but the union classified them as management and thus ineligible for union membership.

As work in the mines became scarcer, there were more men with third-class tickets than jobs for fire bosses. Fewer men were prepared to subject themselves to the gruelling course work and exams. But when the big mines started to close, many experienced miners started smaller operations employing only a handful of men. If there were more than ten of them, the small operations had to employ a fire boss.

> My father says, "Let's try on our own." So we needed a fire boss. So I went to night school. A man named Nichols was crippled in Number One mine, but he had his second-class paper so he took on to teach twelve of us because the company needed fire bosses too. The old ones were dying off and nobody was taking up mining because mining was coming to an end. He was a good teacher and eleven of us passed the exams. He was paid by [Canadian Collieries superintendent] Col. Villiers out of his own pocket.

A second-class ticket made a man a pit boss or overman; a first-class ticket made him eligible to be a mine manager. Very few men were interested in doing the years of study necessary to be a manager, but it was theoretically possible for a youngster of fourteen to begin work on the pithead and end up managing a mine.

One of the small companies working in the 1930s tapped the old Wellington seam north of Nanaimo at Lantzville. During the First World War, a company named Nanoose Collieries had operated a mine in Lantzville "not much bigger than a well." Soon after the war Jack Grant came along and sank a shaft on the shores of Nanoose Bay then sold it to Nanoose Collieries, but people called it Grant's mine until 1920 when a Seattle company bought it, a company whose managing director's name was Lantz.

Lantz's mine was out in the middle of nowhere; its employees had to commute by bus from Nanaimo. So the company built twenty-six houses on the hillside close to the mine. With modern plumbing and plastered walls and ceilings, the houses were very popular and filled quickly. The following year, 1921, the company donated a site for a schoolhouse and social hall and, in 1922, built fourteen more houses and a small ice plant. It was fitting that the town be named after the man who made all this happen.

The coal was hard and of good quality, but lay in two very irregular benches fairly close to the surface. The 2.5-by-5-metre shaft was only 33 metres deep.

> That's not very deep. I could holler from the bottom. These ones in Nanaimo, you couldn't even see daylight.

> They sunk a slope for getting air. Before that they took the mules down on the cage and it took a long time to get them up every day 'cause they didn't have a stable underground like other places. When they made a slope – it's

> right alongside Dickinson Road – you could walk out of the mine and take the animals out that way.
>
> They used to ship the coal by scow. It would come up the shaft and be dumped into a bunker. The scow would come up under the bunker. I think the pilings are still there out in the water. The coal cars could dump right into the scow too, dump down a chute. When the scows were full, they'd phone a tug and it would tow it to Vancouver.

By 1923 the company name had changed to the Nanaimo-Wellington Colliery but the best coal was under the ocean in a seam that pitched gradually upward reducing the already thin cover overhead. The inspector placed restrictions on development.

> They gave 'em permission to go out there and just have one tunnel, like one level, one slope, and just have a single room with single stalls to go off that. Then they'd take a little more off and a little more off – first thing you know they had a great big place with just posts supporting the roof. And they were scared. They didn't know how much rock they had up above the roof. That thing could have broken in and flooded and drowned everybody. Never get any of 'em out. So that's why they shut it down. Some say you could see beach boulders showing through.

The company sold out to Canadian Collieries in 1926 and the new owners filed notice of abandonment. In 1927 the Diamond Jubilee mine with four employees worked in the Lantzville area for a year developing the property near an old prospect tunnel into the upper Wellington seam that outcropped on the shore. Two years later Lantzville Collieries began to search for coal, but it took two years before they found a seam. Throughout the 1930s, however, approximately twenty men work steadily on the basis of co-operative earnings, mining and loading the coal by hand and shipping it on the E & N Railway.

There was steady work but Lantzville was still isolated. On Saturdays, residents depended on the jitney bus to take them to Nanaimo to shop, see a show, play some pool, and wait for the return trip at 9:00 p.m. Occasionally there was a dance at Pleasant Valley, the nearest community between Lantzville and Nanaimo, but most social activity centered around the hotel.

The Lantzville Hotel opened in 1925 to provide a more attractive alternative to a bunkhouse called "The Longwall." Like other hotels in coal camps, the Lantzville Hotel provided room and board to bachelor miners and beer to the thirsty ones, even on Sunday, for those who knew how to do it.

The little store in Lantzville had a room where the miners went to gamble at cards. Gambling debts were marked in a book: "One half dozen five-cent chocolate bars, one five-cent soda pop, one twenty-five-cent bottle near beer . . ." Fortunately for the morality of the town, the wins and the losses usually balanced out.

Lantzville still retains an isolated quality that present-day residents prize. The Lantzville Hotel still dispenses beer and its clientele are mostly regulars. Several former miners settled permanently in the village. Their bar talk was about the five feet of king coal lying out there just under the ocean floor and who was a strikebreaker and what oil company was nosing around asking about coal deposits.

> The ground hasn't changed much. I could show you where the tipple was, where the shaft was, where the powder house was and all that. Only where the shaft was there's a big flower garden now. But they say it still caves in, still sinks down.

The majority of miners who worked in Lantzville at its peak lived in Nanaimo and Northfield, commuting to the mine on the jitney. Jitneys were a common sight on Vancouver Island, but since they were privately owned and each owner provided whatever kind of a vehicle he could find, there were no two jitneys exactly alike.

The two jitneys or whizbangs that travelled to Lantzville were open-air buses with seats on each side and room to stand between the two rows. For twenty-five cents a day return, a man could travel to and from work if he was prepared, wet as he was from the mine, to brave the wind whistling through the open sides. The only alternative was to walk. Which is how the Nanaimo men from Five Acres got to the harbour where the *We Two* waited to take them to Protection mine. "Shank's mare" took them over Miners' Hill and down Albert Street to the dock where the company tug waited.

Miners employed at the Harewood mine in its 1920s incarnation got to work in a coal truck whose owner converted it into a bus.

> It had a bench on each side of it that he'd slide in when he wasn't hauling coal. The benches were longer than the body of the truck and two or three men would be sitting out over the end. There's be ten that was sitting on the inside balancing the three that was on the overhang. An adult would pay $5 a month and a boy paid $3. I was a big boy so I had to pay $4.

Three companies took passengers to the Granby mine. The Bamford line served the highway south of Nanaimo with charobancs, which had crosswise full-width padded seats and a door on each side of each row. Twenty-five or thirty passengers could travel as long as they didn't mind the abundance of fresh air that came through the open sides unimpeded by the wind flaps. The Stafford bus company that travelled the Nanaimo to Granby route had windows like a modern bus. Mackenzie's jitney plied the South Wellington to Granby route and Johnny Sisco's jitney did the same between Granby and Ladysmith. Tommy Simpson, Donachies, and the Main Brothers were other private bus lines on the Island.

Braving roads full of potholes, the private bus lines changed the character of the coal camps. Because they made it possible to live in one place and work in another, it was no longer necessary for a family to move house when a mine closed. The population of each camp became more stable. But the buses made the camps less isolated too and less dependent for their own entertainment and help. It had become possible to live in South Wellington or Extension and spend Saturday night in Ladysmith or visit a hospitalized relative in Nanaimo. Buses made these things possible but they did not impress everyone.

> They'd be waitin' there at the mine at the end of the shift, but it was just the older people that took them. Mind you, I liked my dad to take the bus home because I could run home faster than he could ride and so I got the hot bath water first.

In the days just after the First World War, when the markets were good and the miners' skills in demand, when Morden, Reserve, Granby, and Lantzville were added to the mines that were producing coal and jobs for men eager to put the Big Strike and the war behind them, when bars and dance halls and fancy ladies offered diversion, it was good to be a miner. It was good to be a miner unless he remembered the failure of the union to gain recognition and ignored the circumstances that still put him at the mercy of the boss. For most men, the 1920s were a time to lick wounds and prepare for another battle.

CHAPTER X

We Too

> Even an animal will fight when it has to. The most timid animal if it's cornered will try and fight. And this is what it was about I guess. Life's not worth living if you haven't got a bit of freedom. And this is what we fought for – freedom. – Jock Gilmour

> In the old country the union is the union. When you start to work one thing you must do is join the union. You're listening to the older men and you know why it's there and what it meant. But out here it was a different thing. When I come to Canada I brought my transfer from the Scottish Miners' Union. I had it in my pocket and the first day I went to work I asked a fellow there who happened to be a Scot where the union secretary was. He told me to put the letter back in my pocket and keep my mouth shut. There was no union here. – George Bryce

During the years immediately following the First World War, the western world seethed with labour unrest, but on Vancouver Island there was no apparent union activity whatsoever. Some miners were only recently off the blacklist. Many families were deeply in debt to local stores and would spend years paying it off in small monthly installments. Although new mines were opening and markets were good, a miner could be fired for the slightest hint of union sympathy.

> If the word got out that you were organizin', you were told they didn't want nobody organizing that mine. So you were told to take your tools and you were finished. And with having no union you had no leg to stand on so home you went and you walked around trying to get a job in another mine and if the word got around, you didn't get a job. My dad was let out of a couple of places for talking too much. That's what they called it: Talking too much.

The UMWA had been dead on the Island since 1919, its locals having faded one by one during the war years. In the rest of British Columbia there was a union upsurge, especially among miners and railway workers. The UMWA continued to grow in the Crowsnest Pass, and radical unionism was making its presence felt all over Western Canada. The Winnipeg General Strike in 1919 had frightened the Canadian government, which was convince the strike had been influenced by the recent Bolshevik Revolution in Russia. For years afterward, the government and its law enforcement officers would associate any labour activity with communism and would brand workers "Bolsheviks" – a word some of them had not even heard before. A decade later, the label would be a closer reflection of the truth as the Communist Party cell system became the tool that made it possible to organize the union the second time around.

But at the end of the war, Island miners were, on the surface at least, docile. As they licked their wounds, however, one man's defiance reassured them that some day they would again rise and fight for the union. This man was Ginger Goodwin.

Albert "Ginger" Goodwin was from Cumberland, where people knew him for his abilities as a soccer player and a speaker. Ginger was a union man and had been through the Big Strike. By 1917, he was on the executive of the B.C. Federation of Labour and was the president of District 6 of the International Union of Mine, Mill, and Smelter Workers. When he fell out with that union he went to the Kootenays, organized the Trail Trades and Labor Council, and led a strike for the extension of the eight-hour day. Ginger had been declared unfit for military service because he had a racking tubercular cough and teeth so bad he couldn't chew his food, but after leading a strike in March 1918, he was reclassified as fit to fight in His Majesty's service.

Ginger disappeared. With several other draft evaders he headed back to the Island and holed up in a cabin near Comox Lake not far from Cumberland. Local miners, taking turns to make the trek over land and water, kept the group supplied with food. Then a woman in Cumberland informed on the hidden men, and police infiltrated the woods around the lake. On July 26th, Dominion Constable Dan Campbell shot and killed Goodwin with a soft-nosed bullet.

Sometime in the week following the shooting, an undertaker in Courtenay received a visit from some men who asked him to bury Ginger up at the lake where he lay. The undertaker refused. A Cumberland undertaker received a similar request; his answer was the same. On August 1st, a coroner's jury ruled justifiable homicide. The next day, miners carried Ginger's casket on their shoulders through the streets of Cumberland and out the road to the cemetery.

Earlier that day, August 2nd, the Trades and Labor Council and the Metal Trades Council had called all union men out for twenty-four hours to protest "the shooting of Brother A. Goodwin." By noon, the majority of Vancouver's workers had laid down their tools and the first general strike in B.C. history became a reality. Twenty years later, when the UMWA finally achieved recognition on the Island, a unique stone appeared in the Cumberland cemetery to mark Ginger's grave. The hammer and sickle is incorrectly rendered, but the people who have maintained the grave and painted the outline of the emblem in red and black are sincere and so are the words:

Lest We Forget
Ginger Goodwin
Shot July 26th
1918
A Workers Friend

Vancouver's workers may have walked out for twenty-four hours, but the level of labour activity on the Island was very low. Island miners gave little support to the One Big Union (OBU), which attracted 41,500 members in the rest of Canada in 1919 in its first year of operation with its resolve to form a revolutionary industrial union of all Canadian workers. All the UMWA locals in Alberta and B.C. switched allegiance to the OBU as did the majority of the members of the Vancouver Trades and Labor Council.

The desertion of the UMWA by its locals led to strange alliances. The international headquarters of the union agreed to co-operate with employers to entice straying members back into the union fold. Even the federal government, alarmed by the professed revolutionary nature of the OBU, passed an order-in-council in 1920 requiring that all men working in and around mines and eligible to become UMWA members *must* join the union.

The irony must have hit Island miners with special force. To be required to join a union they had fought for and lost just six years before was laughable, especially when that union no longer existed on the Island. Nor would Island mine owners have tolerated the order-in-council.

A few Island miners responded to the quiet entreaties of OBU organizers, but by 1921 that union was dying, its total membership having declined to 5,300 in just two years. Critics of the union said it was obsessed with structure and that it often forgot the concrete problems of the workers.

> I joined [the OBU] in 1922 because my dad was a union man and I was ready to join any union, but I never went to a meeting or paid any dues before it fizzled.

> There was a union tried to come in, the One Big Union, but it didn't work. I dunno what happened. I believe the head of it ran off to Australia with the money or something.

> My dad had an OBU button and he went out to Morden mine and the boss there said, "Ya better take that button off or you'll never get a job." So my father took the OBU button off and he got a job in Morden mine.

That was the way it was in the early 1920s on Vancouver Island. A man had to choose between the union and being able to work. But some men took out their frustrations in other ways.

> Before we had the union, I was a young man and we had to work five days a week plus some Saturdays if they told you you had to. So one Saturday I didn't work. A young man wants to get out you know. Anyway, Monday morning the old boss stops my lamp. "Where was you at Saturday?" I says, "Oh, I had a toothache." He says, "You were seen downtown." I said, "You had your spotters out did ya?" He says, "You're working next Saturday." I said, "I'll tell you something right now, I won't work." Saturday comes around and I didn't. Monday comes around and I was fired. That's the way things were.

> One of the head men in Nanaimo was organizin' a sports day at Robins Park with horse racin' and bike racin' and lots of sports. So I said to some of the boys, "Why don't we ask the boss if we can get a half day off?" So we went to see him and he said, "Ya, that'll be fine." So when we went to work the next day nobody took a bucket 'cause we expected to be off at noon, but the boss come and he said, "Looks like you guys are gonna work a whole shift." So I just says, "Well, up your..." and I'm down the river. Anyways I didn't care. You could quit a job and get hired somewhere else pretty quick.

By 1924, that was no longer true. A depression hit the markets and a man could no longer find work "pretty quick." There were at least three reasons for the market slump: it had been artificially strong because of the war and now its true nature was showing; fuel oil was cleaner and had become more popular; newer coal-burning machinery used less coal. By the time the Great Depression hit the rest of the economy in 1930, 682,000 tonnes of Vancouver Island coal per year had been displaced. Island mines, equipped and manned to produce 1,136,000 tonnes annually, could no longer sell all that coal.

Lt.-Col. Charles Villiers, general manager of Canadian Collieries (Dunsmuir) Ltd., explained all these factors to a Vancouver meeting of businessmen. Since 1927, when it bought out Western Fuel, Canadian Collieries owned all but the smallest of the Island's coal mines. When men spoke of "The Company" in the years that followed, it was to Canadian Collieries that they referred.

And when men spoke of "The Boss" they meant Colonel Villiers or John Hunt, a man who had parlayed his natural abilities into a long career in coal mine management and one who had both supporters and bitter enemies among the men he employed. Villiers was an ex-British army man, "what we called a 'long-stocking Johnnie'," and he had some supporters too. It was Villiers, the butt of many a cartoon and slogan in the clandestine union newspapers, that the men blamed for the repressive policies of the company.

Pit bosses came in more direct contact with the men and there were a few of them whose treatment of the miners made them notorious. Anyone who worked in the mines was convinced that there was a particular boss who had it in for him.

> Before the unions come here it was a real crime. If you looked cross-eyed at the boss you know what he'd do to you? You'd go in the morning and you couldn't get a lamp. The fella in the lamp cabin'd say your lamp's stopped for two weeks. Well, you're off the payroll for two weeks. They had the horsewhip over you. I know what a union is. It's a big thing for miners. You betcha.

> It wasn't the fourteen-inch seams or layin' in the water. Miners might not have liked the conditions they worked in, but they took their licks and they didn't want somebody comin' to tell them to get a move on every time of the day and shufflin' them around and threatenin' them. It's this continual threat, you know. And somebody could come around and say, "Well the mine is gonna close down." Threaten with closures.

> When I started in the coal mine I thought, this is forever. This job is forever. This mine is gonna last forever. That's how stupid I was at that time.

> Before you were a slave. You couldn't go anywhere, you couldn't do anything, the boss just run you, ruled the roost. If he didn't like you well, he could fire you. For no cause. In South Wellington the mine used to shut down every summer and I remember on one occasion, the notice went up: This Mine Will Work Sunday. Well, them days, if you took a day off you'd get fired. So we worked thirteen days straight and the following weekend a notice went up: This Mine Is Closed Down Indefinitely. That is how they acted. They pushed you around like a dirty shirt.

> Fred and I worked together in Number One and we were almost ready to take the fire boss and destroy him. That's how bad it was. The bosses organized the unions. The men were forced by management. No question about it.

Vancouver Island coal mines were not particularly disagreeable places to work, as coal mines go. The conditions were often better than those in Cape Breton, Pennsylvania or Great Britain. The CMR Act was highly regarded and the B.C. miners' ticket examinations considered tough and thus likely to produce high-calibre graduates. But there was no use having highly regarded legislation and strict regulations if they were ignored. In the same way that the mules were valued over the men, the company gave top priority to the removal of as much coal as possible even if it meant ignoring an injured man.

> Well we were treated like rats. When I cut my finger off I was sitting there from quarter to six that night 'til quarter past ten holding my two fingers. One was off and the other had the nail off. Over four hours to get a ride up.

As long as there was coal there, they pulled the coal. I never had a band-aid or bandage and they never had no place for the first aid stuff. Well we fought for getting a shed built down in the mine for all the first aid stuff.

In Extension mine two men worked together. One was killed in the early afternoon and the one that was still alive come out with the dead man, eh? They docked him that half shift. Now how low can a company get docking a man that half shift for coming up with this dead body?

If you were injured and someone replaced you, you might not get your job back when you were fit. They'd look for someone to replace you and a guy might be waitin' at the mine at startin' time with a jam sandwich in his pocket. In other words, he was ready to start. Somebody would throw him a cap and he'd have your job.

Coal miners in British Columbia had been among the first to win the eight-hour day. But the company had found ways to distort the eight-hour law. Men and boys working on the pithead often worked nine or ten hours but received pay for only eight. Cumberland miners were required to work eight hours bank to bank meaning that their shift did not start until they reached the actual coal face and ended when they left it. Not only did this mean that they had to travel underground great distances, sometimes kilometres, on their own time, but it also meant that two shifts were in the mine at the same time.

Before we got organized we used to go down the mine at 3:00 and the other shift is supposed to be coming up as we're going down but they're still away down in the mine still pulling coal and we're walking down. The company's after all the coal they can get so they're getting two shifts down in the mine at once which wasn't allowed. In case of explosion you kill twice as many men.

In the Nanaimo mines the men worked eight hours portal to portal, meaning they were paid from the time they stepped onto the cage until they arrived back at the pithead. For each cage-load of men who descended the mine, a cage-load was to come up from the previous shift. In this way there would never be more than one shift in the mine at one time.

An eight-hour day did not necessarily mean a five-day week. Saturday and Sunday were often used for company work – replacing damaged timbers, installing new stoppings. A man working on haulage, whose replacement on the next shift didn't show up for work, was expected to go back for his mule and work for another eight hours at the regular rate.

As the Great Depression deepened and the rest of the world succumbed, the company and the miners ignored the violations of the eight-hour law. Men were glad to work two shifts back to back. Markets failed, the economy slowed and unsold coal piled up in a field near the coal wharves.

During the Depression we couldn't sell the coal. We had mountains of it down here, piled up. Then some big orders would come in and they had men down there shovelling it up with pitchforks back into the coal cars. The men were working maybe one or two days a week. We used to have a whistle on old Number One mine. Every supper time around 4:00 or 5:00 your dad would say, "Now everybody be quiet and see if there's work tomorrow." If it blew one blast there'd be work; if it blew two you stayed home the next day. South Wellington had one blast for work and four blasts for no work.

They let the single men go first and the only thing left for them to do was to work for their board. They had them working on the road. They had miners drilling holes with hammer and steel when there were jackhammers and compressors in the mine shed doing nothing. If they'd have given them jackhammers and paid them $4 a day they could have built all the roads they ever needed.

When we used to be lookin' fer a job, I've seen as many as thirty and forty men sittin' on the old fence outside the office and the boss'd look out the winder and he might call one man in there for a job. And that was week after week they done that. They'd say, "Come back tomorrow." That's what we called Jack Hunt's assistant. "Come back Pete."

Those were tough times. Miners got only two or three days a week at $2.25 a day and the city sometimes gave us some work. Just depends on 'ow many family you 'ad. The bigger the family the more days you worked. They gave you scrip for food instead of pay. But them days you was glad to go and get two or three days' work. And if you spent two bits to go any place you were short on the table. The government brought out relief after 1932. You 'ad to work on the roads. We got $17.50 a month. On $17.50 a month you don't buy no fur coats.

The residents of the coal towns had known hard times. They shared garden surplus, caught fish and hunted game; neighbours watched out for each other.

Miners received coal at a very low rate. Other people did not, so when hard times came, you saw to it that they never went cold. In those times we shared what we had. There was many a time that I carried a bucket of coal to our neighbours or somebody did. Always at night. You left it on the back porch.

We got a special permit from the game warden and I used to kill deer for all the people. Then there were two other fellas with a horse and buggy and they used to go around collecting vegetables off the farmers. And we'd fish for salmon and trade some of it for a loaf of bread. That's what we hadda do.

Children born during the Great Depression were unaware that there was anything unusual about the way they lived. They just got on with being children. They played soccer, softball, and lacrosse; they swam and played cards. The stakes in the card games were high: wooden Eddy's matches swiped from Mother's big box over the stove. If a kid could muster up the money, he could see a Tarzan serial, a travelogue, *The News of the World*, and a feature movie all for a dime.

But to afford a movie someone in the family had to be bringing in some money. Timberland mine, south of Extension, had seemed to offer hope for the jobless when it opened in 1928 – shortlived hope as it turned out. The mine closed after two years. But about this time a number of miners took fate into their own hands. Following the Richardsons' and the Fiddicks' lead, several small operators went into business for themselves.

In the heyday of coal mining, the big companies made it difficult for small operators to stay in business. There were a few families with coal rights, but it was hard for them to find a way to ship the coal they mined. The owner of the Jordan mine near Brannan Lake built a wooden railroad from the lake to Nanoose Bay. Big wooden cars pulled by eight-mule trains carried coal over wooden rails. The railroad delivered three shipments that way before the mine went out of business. Thirty years later, Ralph Chambers and Pete Ross made a lot of money out of the same mine – this time transporting the coal by truck.

There was a local market for heating coal and some – the Biggs, Bill Loudon – managed to make a living selling it. Chamber's mine on the Old Number One Extension site and Beban's mine right beside it employed a number of men and pulled a fair amount of coal. The equipment in the smallest mines was makeshift – car engines ran the ventilation fans, buckets hauled out the water – but the mines passed inspection so the makeshift measures must have done the job. The small mines worked out well for those who were interested in trading hard labour for a little money and a lot of independence.

> If you couldn't find a job you made a job. It was that way because the big mines was finishing. And we were still in business ten, twenty years after the big mines finished. And we had all the market then, what was left.

> We didn't make no money. We were all partners in it so if one didn't make anything, the whole works of us didn't make nothin'. There was lots of times when we hit the old workings and we wouldn't get no coal for a month maybe. We'd have to keep goin' until we found these different stumps that was left. There were big openings in there where they had been mined out before. It was slow workin' through them.

> Two of us had third-class tickets and we applied for a lease on some company property. We worked there seven or eight years. Workin' for ourselves, of course. We paid so much a ton to the company or the government, I forget which, but it was a nice life. You were independent. If you felt like quittin' for the day, you went home, that's all.

My wife used to worry because I was by myself. I was at my last mine and my partner died. The inspector shouldn't have allowed me to mine alone but I knew him well and I told him, "I only want to finish this piece of ground off and I'll be finished in six months." So I finished the mine off, didn't have any trouble at all. It was a nice piece of coal and when it finished I stopped.

But for every man who found satisfaction in digging coal for himself, there were many who did not find the coal they wanted. Many more got through the Great Depression by working for the company, when it had coal to move, and working for the government on relief gangs when it didn't. And there was competition for those non-jobs too. The warm West Coast had become a mecca for the poor from the Prairies who would work for even less than the local people would.

When the dust hit Saskatchewan and that, they were bringing those poor people here to the coast on flat cars and dumping them off here on the Island and these poor people were so hard up and we were working for 20 cents an hour, eh? And they were coming down offering to work for less than that.

But relief was not an adequate replacement for employment and there were many idle days. The government began to send single men, who did not qualify for relief, to camps to reduce the chance of them causing trouble. A meeting in Ladysmith to demand relief for single men instead of sending them to camps included some men who had walked or hitched from Victoria. It was pouring rain when the meeting ended.

After they finished talking they were standing around in the rain thinking about how to get home. One guy says, "Get the train." He meant ridin' the blind, that platform where the conductor couldn't see you, but when the passenger train arrives from Victoria they all get on and sit in the coaches. When the conductor comes along to get the two-bit fare, nobody had any money. So he took all their names and a month or so later they were all summoned to appear in court. They all got a year's suspended sentence for riding on CPR property for nothing.

Meetings of the unemployed always drew a large crowd. The ones at Richard's Hall in South Wellington filled the place with people drawn by the promise of shoes for their children. Whole families would march on Ladysmith where the local provincial policeman also served as the relief officer. What many of the participants didn't know was that the Workers' Unity League had organized the marches. The purpose of the recently founded arm of the Communist Party of Canada was to organize miners, loggers, longshoremen, and the unemployed.

The will to organize and fight back was alive again on the Island. It had made a tentative appearance several years before in 1925 when a short strike halted work over a decision by the company to lower wages. They lowered the wages anyway, but the workers had some say in how it was done. A small start indeed, but evidence that the men were ready again to take united action and stick by it.

> In a way we kind of organized ourselves. In the twenties sometime, we were paid so much a day and we decided that twelve cars was plenty for one day. They were two-ton cars. The company said they were ton-and-a-half, but we measured them and they were two-ton. Anyway, two-ton, ton-and-a-half, they were still getting at least eighteen ton for $6. That was the day wage. That's nine ton a man. So we wouldn't load any more.

Management had seen action like this before. When times were bad, the solution was simple. Fire the troublemakers and replace them with men who were easier to get along with. But the miners had learned a few tricks.

> The boss would know he could get after the softer ones who didn't have any guts. They'd start loadin' extra cars, but when they needed help putting up an eighteen-foot timber in those high places in South Wellington, it was a hard lift even for two men. Well there'd be no one to help 'em. So they struggled with that timber and sometimes they couldn't even fill their twelve cars and they pretty near got fired. So that's how we did it.

As the Great Depression loomed and work became scarcer, more and more such incidents took place. Localized and informal, they showed a readiness for the return of the union.

> We were deliberately going slow and the boss, who was a Scot, was trying to catch us in the act, but he never quite did. We'd tap on the compressed air pipes to let each other know where he was. One day he come in and he held up his safety lamp and he said, "Now I'm not gonna fight wi' ye, but as sure as this lamp the mine'll shut doon. There's lots of hard coal in Cumberland." "Well," I said, "that's too bad. I'll have to go where the hard coal is. I've got nothing to lose."

In 1927, the Mine Workers Union of Canada (MWUC), an affiliate of the All-Canada Congress of Labour (ACCL), had designated Vancouver Island as District Two and made preparations to organize locals in the mining camps. The organizers had the field to themselves, but they were up against an atmosphere of fear made more acute by the decline in the number of jobs. Enough men in Cumberland were willing to risk forming a local in 1930, and they launched a series of small strikes. But the strikes accomplished nothing. It seemed as if this union effort would fail too.

Then the Communist Party appeared on the scene. In 1929, on orders from the Comintern – the international association of world communism – the Canadian branch of the party began an experiment in union organization under the name of the Workers' Unity League (WUL). Although short-lived, the WUL proved very effective, using as it did the underground cell system. The MWUC dropped its association with the ACCL and joined the WUL, bringing in its Cumberland local in 1931. Sometime in the next year, the MWUC moved into South Wellington.

> Dad was the first secretary underground of the Mine Workers' Union of Canada. The meeting to form the first local in South Wellington was held at our house. I remember that meeting very plainly. We were all supposed to go to the dance and Tommy Shore, the district organizer, and Ed Pearce drove up at 5:00 and they said they were going to have a little meeting. The dance started at 9:00 and Mum and me waited until 11:00 and they were still meeting so we went off to the dance by ourselves. Mother was so mad. She didn't know what the meeting was all about. She was so mad she could have shot Dad.

It was not by accident that Cumberland and South Wellington were the first camps to organize. Both were small communities entirely dependent on coal for existence. Extension was in a similar situation, but because the mines had closed there in 1931, the union-minded citizenry supported organizing wherever they were working. Miners in Nanaimo were more reluctant to support a new drive towards unionization. The town was larger, its population more diverse; it did not have the unity of purpose that the smaller camps had. The main employer, Western Fuel, although now owned by Canadian Collieries, was much harder to deal with. Organizers could only speculate on the reasons for this: Did management discourage unions more vigorously? Were there more informers? Were conditions better at Number One?

The strategy proposed by the WUL organizers, who by this time were travelling extensively up and down the Island and contacting miners, loggers, and longshoremen, was the use of the cell system.

> There were so many in a group, maybe eight, nine or ten men with a group leader at the head. By only having a small group you did not know who was in the other ones. If you were a friend of the bosses, you couldn't tell the boss, "So and so belongs to the union." Then you had your little group meetings and you know it was very scary when I think back now. You were really afraid for your job.

> People wanted the union. The ones that came to organize them lived on next to nothing. They never got no money for it. There'd be meetings in somebody's basement or they'd be out in the bush and someone would take a couple of loaves of bread for them. Another one would maybe be out fishing and take 'em a couple of salmon. Some of them were good speakers. One lad, I forget his name now, he coulda started a revolution if he put his mind to it.

When the federal government declared the Communist Party illegal in 1931, secrecy was even more necessary. The work of organizing began with the men who were committed to the cause, the men who had been union men in 1912-1914, the men from homes where union talk was part of the daily fare, whose sons grew up on the sure and indisputable truth that the union would rise again. From that core of committed men the movement branched out, each man listening to conversations at work or in the pubs or on the roads, listening for the right moment to approach and quietly talk union.

> We'd take a walk along the track there and form a group, have a meeting or we would go to someone's house and have a meeting. You'd talk to your buddy or the man working next to you and find out which way he was leaning.
>
> Each group had a leader. The leaders would meet in some secret place. One guy used to allow us to use his barn. And you would meet and make certain decisions and then you would carry it back to your group. Then the leaders would take the members' thoughts back to the executive. As a group leader I would know the other leaders but I wouldn't know who was in their groups. This was a very slow and tedious way of doing it, but for beginners it was the only safe way. You also knew that if one of your group got fired and there was no good reason, then someone in the group must have squealed and you could isolate that to one group.
>
> There was ten in our group so if anybody decided to scream to the company about it all they would know about is ten. I was working with the wife's cousin. Me and him were together for years and years. We both belonged to the union but I never told him and he never told me.
>
> At first we had little meetings in Tommy Armstrong's store. He had a candy and tobacco store under the Eagle's Hall there in Cumberland. We had to go very, very slow because you couldn't trust anyone.
>
> When it was under cover they used to send these doggone stool pigeons out to check on everything to see who the union men were. And they'd report back to the superintendent and the pit bosses.

If the cell system in Cumberland and South Wellington was foiling the company and its stool pigeons, the importation of cheap labour was making it easier for management to keep their workforce union-free. The unemployed from the rest of Canada and new immigrants from Yugoslavia may not have been experienced miners, but they were desperate to work.

> These Yugoslavs were great big guys. They could load twice as much coal as a little guy and the company liked that. They were brought over here by contractors or tyees, as they were called, people of their own race, and put straight into the mine despite lots of local men being out of work. They couldn't speak English and didn't realize for a while what the situation was. You should have seen how the contractors had them living. Down there on Haliburton Street they had houses with maybe twelve guys in each house and

> they were double and triple shifting the beds so everyone had a place to sleep.
>
> They were working for about two and a half a day. The regular guys were getting $4.50. They had what they called a pusher to keep them in line. He got them a place to live. If they disobeyed they lost their jobs. They used these immigrants to pressure the regular guys.
>
> When the Slavs figured out what was going on they proved to be more militant than some of the people who had been here for years. They'd been pushed around before in war time and they weren't going to take any more.

More men for the union. The membership grew slowly, held back by the very need for secrecy that kept it alive. Perhaps the most surprising recruits were some of the men who had been strikebreakers in the Big Strike. Many of them were among the most militant in the second stage of the battle.

With some minor exceptions, the women were excluded. Some organizers said they talked too much and could not be trusted. "Telegram, telephone, tell-a-woman," was what they used to say. Even activist Ellen Bowater Greenwell said, "In those days unions were all men's business." There was no women's auxiliary to the union in Nanaimo and South Wellington until after official recognition of the union, and the one in Cumberland suffered from low attendance at its meetings.

But coal miners' wives were not traditionally so passive. Women had added a unique female presence to the riots and courtroom proceedings during the Big Strike. The Pinkerton spy and the local police wrung their hands over the "amazons" and their use of strong language. Less pugnacious women were no less convincing as they supported their men in the fight and struggled to provide the bare necessities for their families. When this struggle returned in the Great Depression, it was the women and not the men who made headlines in the Vancouver papers by demanding action that would develop markets for coal. An outside observer credited the women with maintaining morale in Nanaimo.

It was said that women could not be trusted to keep a secret, but the fiancée of the editor of the *We Too* assisted in its printing and distribution. She was certainly a trustworthy woman. And the women who fed and housed the organizers must surely have known the identity and guessed the mission of these men who appeared without warning, held clandestine meetings and then disappeared. The union organizers were fooling themselves if they thought they were fooling the women.

When Bill Atkinson went to Cumberland in 1934 in search of a job he was not an organizer, but he was interested in unions. His childhood in South Wellington had been steeped in talk of the UMWA. In 1934, however, Bill just needed a job and friends in Cumberland said they could get him one, especially since he could play the trombone and the piano and they needed someone for the band. The Cumberland friends were also working underground for the MWUC and it did not take much encouragement on their part for him to get involved.

The Cumberland local was doing well. Despite the fact that the Communist Party had shut down the WUL that year and had turned its locals back to the ACCL and the UMWA, the WUL's legacy of cell organization had taken a firm hold in Cumberland and the men were ready for action.

> We were pushin' it a bit 'cause the Communist Party was illegal at that time. Then we changed back to the Mine Workers' Union of Canada. Then the company started a union – the home locals we called 'em or the cellar gang 'cause they met in the company cellar. We'd 've been canned if they found out we were organizin' our own union, if they'd found out who was doin' the paintin' on the buildings and fences and all the cars in the mine. "Join the Union and Be One of Us" and all this and that.

> They used to publish a sort of little bulletin. In Cumberland it was called *The Tipple* and in Nanaimo it was called the *We Too*. We used to distribute that. Scatter them out amongst the men. And they would be nailed up in a car that goes into the mine and it would circulate all over the mine so everybody sees it sometime. It was mimeographed.

The Tipple circulated the union view and provided a forum for complaint. Letters to the editor were numerous but signed with pen names for anonymity. Cumberland miners were getting braver, however, as their numbers grew. At a mass meeting in the Band Hall on November 8, 1934, the men on haulage refused to work until their wages increased by 7 cents a day. The mines shut down. Two weeks later, the company and a recently elected grievance committee that just happened to consist of union members signed a two-year agreement ending the strike.

May Day parades were in style again, the 1935 edition being a particularly successful one, much to the dismay of more conservative Cumberland citizens who felt it detracted from the 24th of May festivities. The MWUC marched in the parade for all to see showing that the union was ready to risk identification of individual members. Canadian Collieries retaliated that summer by laying off forty-five men, all union members and including the entire grievance committee. Sam English, local MWUC president, Shakey Robertson, editor of *The Tipple*, Ben Horbury and Bert Davis were among the group that came to be known as the "discriminated forty-five."

Yugoslavian contract labour from Nanaimo replaced the forty-five, but by November, the union was ready to retaliate. Union members walked out, closing Number Five – the only mine in operation at the time – for six months. Now the company knew for certain which men were union members.

> They brought all these guys from Nanaimo up and kept hiring them. Big Slavs you know. They put them in our places and there were too many of us and our union boys were still out so we had a meeting and decided that we ain't goin' to work neither until they hired our men back.

> One thing they did have in Cumberland was a good provincial policeman. There were a few odd skirmishes around town you know and they wanted to get some extra police in and he seemed to know that as soon as the specials come in they start pushing people around. The miners wouldn't take that and there would be a go at it. So he come up to the union hall and says, "I wish you'd keep it quiet. Just cool it down. There's twenty specials in Courtenay and I told them in Victoria if they push specials into Cumberland I'm finished. If you guys keep it cool they'd stay in Courtenay." So there was no violence here.

In May 1936, a conciliation board found against the MWUC, calling them Communist agitators, but the strike ended when Col. Villiers promised to hire the discriminated forty-five back as jobs became available. The MWUC had a paid-up membership of 350, their meetings were public, and in the summer they would go to a meeting in Calgary and vote to join the UMWA. When they negotiated their new contract in October 1936, they did so as union men.

The decision to go with the UMWA was a natural one. When the WUL disbanded it had advised its members to affiliate with the powerful international union, whose dynamic leader, John L. Lewis, was an inspirational figure to many. For the Cumberland miners, the UMWA was a natural choice, it being the union they had fought for in the Big Strike.

In 1936, Bill Atkinson, by then a union organizer, took on a new job: organizing the miners in Nanaimo. He was a natural choice for the job because he was not only committed and popular, with a "gift for the gab," but he was an insider bred in South Wellington. His presence in the area would not give the union away. In Nanaimo, it was still necessary for organizers to be secretive.

> I can remember a meeting in the Finn Hall, which we held on a Saturday night when they didn't expect anybody to be holding a meeting. All the blinds were down and we kept our voices low. Everyone was discouraged and the "weak sisters" were falling by the wayside.

The Nanaimo equivalent of *The Tipple* was the *We Too*, a word play on the name of the tug *We Two* that towed the scow to Protection mine. The *We Too* spared no one who, in the opinion of the editor, did not want what was best for the men. Long-held grievances became legitimate when they appeared in print. The editor singled out bosses and took them to task for their methods in dealing with the men; he exposed management's attempts to divide the men among themselves; the cartoons were merciless.

> And so we used to distribute the *We Too*. Scatter them out amongst the men. And they would be maybe nailed up in a car that goes down the mine. Then there'd be all kinds of things appear like maybe on safe practices or any other thing that happened that wasn't good for the workers. It was mimeographed but who by, that was all secret. I couldn't tell you that even sitting right here.

> Jack Atkinson was the leader of a cell in Ladysmith and he didn't even know who put out the *We Too*. Tommy Lawrence was the editor. He got deported later in 1941, but he was the MWUC organizer on Vancouver Island and I used to deliver them around on my bike to the cell leaders.

> *We Too*, Vol. 1, No. 3, March 23, 1935.
> The bosses are not interested in the welfare of the men. The more they are driven, the more they are speeded up, the more profit goes into the bosses' pockets. While you are riding backward and forward in the [*We Two*], think over these things, take note of the time she leaves this side, the "comfort" we have while travelling. At the end of the shift look around at your fellow miners sitting around the scow cold, wet and miserable. The day's work is finished; the company has ceased to be in a hurry as far as you are concerned. Think of what a force all these miners of Nanaimo organized into one solid body would be. Then the scow would leave Protection side at twenty minutes after the hour.

Cell leaders meeting in the Finn Hall decided to invite the UMWA to come in and take over. Tom McEwen, the MWUC secretary in Vancouver, asked men like Archie Greenwell to prepare the cell groups for the transition. One of the UMWA organizers coming in to help would be Bill Atkinson.

> Bill Atkinson and Tony Radford they come and they told us what was goin' on and what they were going to do and asked us would we sign up? "Definitely," I says, "I sure will." So anyhow, I says to my kid brother, "What do you think? Let's go. If we lose our jobs, we lose our jobs. It's impossible to lose. If the biggest majority will turn around and sign the union thing, how are they going to fire us?"

> The UMWA came in with their huge organization and money and they come right out in the open and called open meetings.

> A UMWA organizer from District 18 in Calgary called a meeting in the Eagle Hall basement. There was a good turnout but not many people signed. So as they were going out of the hall the organizer says, "There's management spies outside. They'll see you coming out of this meeting and know

what you've been up to. You'd better sign now or you'll be out of a job." So a lot of them went back in and signed.

We come and we told Mum we'd joined the union. "My God," she says, "what's going to happen if you two boys lose your jobs? What are we going to do?" I says, "Mum," I says, "how are we going to lose our jobs? We've done nothing wrong. And if they're going to fire anybody they've got to fire the whole darn shooting match." It would be a different thing if it wasn't the majority, but we all joined.

We all joined up, every one of us and the officials called us a bunch of names but we just ignored them. We seen where it was coming from. We figured the time will come when this will be all thrashed out for our benefit and their benefit.

Then came the day we thought we had enough strength to declare ourselves. Me and my pal, we were supposed to go to work on Monday morning and wear our union buttons. So Monday comes along, we puts our buttons on, and away we goes to work. And we were the only two that had our buttons on. We had our dates mixed up. Well being young and foolish, we were too stubborn to take them off. That afternoon the mine manager came by and had a good long look as he passed me and he said to the next man, "What the hell is that guy wearing?" I never heard the answer but nothing was done. The next week everyone wore their buttons and that was it. At that time there were 600 underground between Number One and Protection and I'd say a good two-thirds of them were union men.

You were supposed to pin your badge on your coat or on your shirt so it showed, but some were scared and they pinned their badge underneath the collar or something.

We come out in the open. We thought we had 50 percent of the men and by golly the company gave in. They recognized the union, signed the check-off, and when we got the lists we only had 30 percent.

They didn't go after wages or anything. They just kept everything running smooth and I think the companies thought the unions is wonderful and the companies is wonderful. If it wasn't for the company, you wouldn't need the unions.

We figured we needed the international union. Than you can deal with international companies and you have some power. The companies finally come to realize it would be better for everybody.

See, I guess what they figured, what they had in their mind was that we'd get organized and just want to slough off and we'd care less. But we were not that kind of men. We was out there to do our day's work and do it in a proper manner and produce. I wasn't going to go down there and work for $5 or $6 a day and not earn the money.

> Before we had the union we was *told* to do everything; after we got organized, they *asked* us if we'd do it.

In the spring of 1937, the Nanaimo and Cumberland union executives and the District 18 UMWA representative confronted the company with the information that 50 percent of the 4,400 miners in the Nanaimo area had joined the union. The men wore their union buttons in full view and the company conceded defeat. The agreement signed by the two parties equalized daily wages: the $6-a-day men came down and the $4.50-a-day men went up so that everyone would receive $5.50 a day. The agreement went into effect on June 15, 1937. In January 1938, Beban and Chamber's mines signed, making Vancouver Island 100 percent organized for the first time.

On November 18, 1938, John Hunt, general manager of the Western Fuel Corporation and H.R. Plummer, successor to the recently deceased Lt.-Col. Villiers as general manager of Canadian Collieries (Dunsmuir) Ltd., joined Dick Cole, president of the Cumberland UMWA local, and Ed Webb, president of the Nanaimo UMWA local, and other local and District 18 representatives in the official signing of the agreement that had already been in effect for eighteen months. A photographer preserved the occasion for posterity. Six weeks before, Number One mine in Nanaimo had pulled its last carload of coal to the surface.

The biggest mine on the Island had shut down, but there were new reasons to be optimistic about the future of coal. No one took John Hunt seriously when he gloomily predicted that "grass would grow on the streets of Nanaimo." Markets had begun to improve in 1936 as the world began to take Hitler seriously. And as if to confirm that coal mining would go on forever, a new mine opened at Northfield and suddenly there were jobs again.

Northfield was not really a new mine, but rather an abandoned Nanaimo Coal Company mine that had tapped the part of the coal lease adjacent to Robert Dunsmuir and Sons property at Diver Lake. All the coal belonged now to the same company, Canadian Collieries. Northfield mine may have provided a lot of jobs, but it was not a popular mine to work in. Its rotten shale roof was hard to support, and its low seam made for cramped working conditions.

> When they first opened it up it was good. It was say three feet, three and a half feet and then we kept going and we got into this lower seam. I believe there was only sixteen inches. We had sixteen-inch posts with two-inch lagging on top. That brought it to eighteen inches. Some places you got hooked up to the roof. Lucky if we had any pants left. And it was wet, filthy, dirty, stinky. It was sort of sulfur and it sure smelled rotten.

I was working on haulage in Northfield and the time I really decided to get out was when I worked in the top seam. When I started on the top seam I could stand up and never hit my head at all. Later on when the mine started squeezin' it was so low that I couldn't get my fingers between the roof and the top of the car and every time I took a car in there, two men had to notch the posts to get me back out. When a mine's squeezin' the stringers can get flattened down as thin as a plate.

What most miners remember about Northfield mine was the day in June 1937 when it flooded. It flooded for the same reason the mine at South Wellington had flooded in 1915: surveyors' maps of two adjacent mines were drawn to different scales. When the Dunsmuirs abandoned the Wellington mines forty years before, old Number Five mine had filled with water. It was theoretically possible to plot the position of the old tunnels to make sure the Northfield workings didn't come too near to them. Then the company decided to go after the remaining coal in old Number Five. Keeping in mind the danger of a sudden release of water, the plan was to tap the old workings in a controlled fashion to allow for the gradual draining away of the trapped water, which would then run through the new mine to a sump at the base of the shaft. From there it would be a simple procedure to pump the water to ground level.

The old surveyors' charts told the miners they were one hundred metres from the nearest workings. In reality, only two to three metres of rock lay between them and old Number Five – not enough to hold back the weight of the stagnant water.

The morning shift had drilled the holes, fired some nineteen or twenty shots and the afternoon shift was to come and clean out the stuff, see. By the time we come down into our places this six-foot wall of rock was getting weaker and weaker and out she comes. An enormous amount of water there. Oh boy, I'll never forget it.

I happened to look along the wall and I could see something a way back there, like somebody was waving a light. I could tell there was something wrong because the air wasn't working right. It was reversing itself. I hollered along the wall to my brother and he hollered to the next fellow and the superintendent on top of the wall there yells, "Come on fellows, get out." We all left everything the way it was and away we goes.

Boy oh boy. The superintendent says to us he says, "Now listen. You've got to watch what you're doing. There's an awful lot of water coming and you gotta try to get up the best way you can. Try and hang on to one another and we'll march up there, By this time the water was at waist level and eighteen-inch timbers was goin' by us like logs on a river.

In the meantime, all the horses and mules had to be taken out first. Animals out first and then the men follow. They'll always look after the animals 'cause they cost more money than the men. We can't let 'em drown, eh? We got to get them out safe.

For once the horses and mules never hesitated to get on the cage and go to the top. They sensed something was wrong. You didn't even have to put the safety bars on. Those animals walked on there just like a man and up they went.

Then they started taking up the men, nine to a cage. There must have been fifty or sixty men waiting. And they kept hoisting – it was about 290 feet, I think – and if there was any prayers said, we said them that time. We were praying to God nothing would happen to that hoist.

It sure didn't take long for the water to fill in. And it started backing up towards the shaft and my God it was getting higher and higher and higher. I was on the last cage. There were six of us left and the water was up to our necks by now. One little guy was a way shorter than anyone else and we had to lift him up so his head was above water. I thought, oh boy, now we've really had it. If anything goes wrong with that hoist, that's it.

Finally it was our turn. As soon as we got on the cage we had to push a button to let the engineer knew we were ready. Boy I couldn't touch that button fast enough. "Come on, bub. Get 'er going."

Nobody was lost that time. We were lucky. The mine shut down for about two or three weeks until they got all the water pumped out and then we all went back down.

Several kilometres to the south, Frank Beban, a local lumberman, had opened a small coal mine in 1935 to tap a body of coal left behind when James Dunsmuir closed the original Extension mine following deliberate flooding to extinguish a fire in 1901. The new mine was a small operation producing only about forty tonnes of coal a day, but it was one of the steadier ones and provided a welcome source of income to the small group of men who worked there. Because a horseshoe of old working surrounded the mine on three sides, the direction of the main slope had to be plotted carefully if it was to miss the workings of old Number One.

On June 11, 1937, ten men were working as the day shift drew to a close. Nels Shepherd and Louis Tognello worked at the face advancing the main slope. Joe Carr had finished for the day so he gathered his tools and headed up the slope for the surface. Then, realizing that he had forgotten something, he left his tools part way up the slope, and retraced his steps. The rope rider, Nels Shepherd's nephew, Joe, had just ridden the last empty trip of the day down to the face for loading. Just inside a crosscut that left the main slope and inclined away to the side, Johnny Senini worked alone.

Suddenly Johnny heard a rumble and rushed into the main slope to investigate. A wall of water was rushing up the main slope towards him. Someone had breached the old mine. By this time, the five other men on the day shift had ridden the crest of the flood and were safe above the reach of the water.

Johnny was not so fortunate, but as the water hit him he managed to hold on to the entrance of the crosscut. The first person he saw was Louis Tognello, his chest and hand injured from the force of the water that battered him from one side of the slope to the other and back again. Johnny grabbed his arm and pulled him into the crosscut. Then Nels and Joe Shepherd materialized, clinging to a floating timber. Johnny reached out again, but the water was slimy. A sudden surge wrested them from his grasp and swept them up the crosscut beyond him.

Because they knew that the crosscut rose at an angle, Johnny and Louis allowed the water to carry them along as well to what they hoped would be an air space. Unable to see, the water having made their lamps useless; submerged to their chins; and their heads crammed against the stringers that supported the roof, the two men surrendered to the force of the water until they reached a dry ledge out of reach of the water.

> The water was pretty near up to their head for a while and then they spotted a little place where there was no water and they got up there. My uncle Johnny took off his clothes and he rinch 'em and he told his partner, "Take off your clothes and rinch 'em" His partner was a little hurt and bleeding. Then Johnny took a pick and started to pretend to work to keep himself warm.

Just beyond reach, the bodies of the two Shepherd men bobbed up and down in the lapping water. At 3:30 p.m., two and a half hours after the water had broken in, a rescue team made its way down the main slope to the thirty-three-metre level where the flood appeared to have reached its limit. Joe Carr's tools sat high and dry just above the water line, their owner nowhere to be seen. Suddenly the rescuers heard a tapping. They tapped an answer. Tapping again. Somewhere behind a wall of coal and rock, Johnny Senini was using a technique as old as subterranean mining itself to tell his would-be rescuers he was alive. The rescuers sent word to the top and pumping commenced.

Inside the little chamber in the crosscut, Louis Tognello was giving in to despair. He worried about his family, trying to imagine their lives without him. His companion became impatient. Johnny Senini was a practical man. "Forget your family. They're dry. Think about ourselves, how we got to save ourselves."

Johnny's tapping wasn't getting a response anymore. They couldn't keep track of time and both knew that the air in the little space was limited. It was hard not to give in to despair. The dark, the cold, the quiet broken only by the lapping of the water at their rocky ledge, the two bodies floating nearby – then the sound of the pumps penetrated the tiny refuge: the best thing either man had ever heard.

Outside, as the day gave way to a long summer's evening, a crowd clustered around the mine portal: rescue officials with breathing apparatus; miners assembling an extra pump hurriedly imported; young Dr. Lorenzo Giovando waiting helplessly, his skills useless unless the men could be reached; relatives and friends from South Wellington.

By 7:00 p.m. the water had receded enough for the rescuers to see the roof of the crosscut. Using lagging, they improvised a boardwalk just above the level of the water and less than a metre below the roof stringers. As they bellied their way along the boardwalk they heard shouts. Johnny and Louis had seen their lights and heard their voices. It would not be long now.

Then the rescuers met an obstacle. An extra-large crossbeam blocked their way. Beyond the beam, fifteen or twenty metres further in, the two men waited out of reach. Fresh air was now filling the little space, but Johnny and Louis had been cold and wet since 1:00 p.m. There was nothing to do but wait and hope they would last until the water receded below the level of the crossbeam. Centimetre by agonizing centimetre the water level dropped.

At 10:00 p.m., nine hours after the floodwaters had invaded the mine, the crowd of watchers saw figures emerge. First came rescuers carrying two stretchers bearing the bodies of Nels and Joe Shepherd – Nels the father of four children under five years of age; Joe married only six months. Joe Carr's wife and baby would wait until the next morning before his fellow miners found his body in a little stall to the left of the face where he had sought refuge.

Following the bodies of the dead miners, the crowd could see someone helping Louis Tognello to an ambulance. Then Johnny Senini appeared at the mine portal, determined to walk unassisted, jaunting and joking, uttering the phrase that would make him a legend: "Give me a Scotch."

> And that uncle of mine he come out and they had the police and the ambulance and they took him in there and everybody wants to be near so I can't get near him. Then he called me to come over and he said he was feeling all right and he said he could eat a steak right then. He said what he wants was a steak. They keep him in hospital one night and then next day as soon as he comes out of hospital he went to the beer parlour and he got drunk – paralyzed drunk.

> He was a tough old guy, good-natured, full of devilment. Never married. He was a great believer in the good in beer. He always said if he was sick he could

go to town, get drunk, and come home cured. Even after he was in the flood for nine hours, the first day they opened the mine, he went back to work. He said it's just a job and we all have to work for a living.

After that we called him John the Baptist.

The spring of 1937 was an eventful one. Four days after Johnny Senini emerged unscathed from the Beban mine flood, Canadian Collieries (Dunsmuir) Ltd. and the Western Fuel Corporation bowed to the demands of the UMWA. It was generally agreed that the union would be a good thing for everybody. With a set procedure to settle disputes, the mines would work more efficiently, safety committees would ensure a better environment, and both the men and the company would know where the other stood with regard to wages and company work.

There were a few holdouts – men who for any number of individual reasons did not choose to join the union. The union men, by now past masters at subtle pressure tactics, set about the "recruitment" of these reluctant men.

You could always tell who the non-union guy was in the beer parlour. He'd be sittin' all alone. Nobody would sit with him.

When I worked in Nanaimo the man that didn't belong to the union you wouldn't get in the cage with him. Finally the manager told him, "We can't run the cage up and down for one man, you gotta get in that union."

For the men who did join the union and were working steadily there were two weeks paid holiday for the first time in the history of the mines. Men who had grabbed bites of their sandwiches between shots and gulped their cold tea as they waited for an empty car, could now sit down for twenty minutes at noon when the machinery stopped for the lunch break. There was a pension plan if a man contributed every month until he was sixty-five. The closing of the mines got in the way of a lot of men's regular contributions, but those who managed to keep up received monthly UMWA cheques for $220 that eased their old age considerably. It paid to be a lifelong union man.

For the organizers who made the union a reality on the Island, there were new fields to conquer. These were unusual men, prepared to live in hiding, to move from town to town one step ahead of the police; these were not men to settle down and enjoy a quiet job and the fruits of their labours. Tommy Lawrence, editor of the *We Too*, was deported in 1941. Tom McEwen, the MWUC secretary in Vancouver, had already been to jail for being a communist. Joe Armitage joined the MacKenzie-Papineau battalion of the International Brigades and died near Madrid during the Spanish Civil War. Bill Atkinson received a promotion with the UMWA and a new job as western representative at headquarters. The Saturday night before he was to leave for Calgary he died in a car accident.

As new markets and new mines provided income to men given new dignity with the recognition of the union, events in the world jostled with each other for space on the front page of history. Adolf Hitler and Generalissimo Franco vied for headlines with King Edward and Mrs. Simpson. Japan rattled her sabres across the Pacific and Mussolini thumped his chest and proclaimed a new Roman empire. Governments conferred and munitions czars rubbed their palms in glee and coal was in demand once again.

The mine whistle had marked Remembrance Day each November 11th morning since 1918, but its dreary message was forgotten as the world geared up to fight what everyone said would be a just war. And like the war to end all wars that ended in 1918, the just war required of coal miners that they stay home and serve their country in the mine.

> All of a sudden there was money available for everything. The miners went from little work and being kicked around to being union men who were needed in the mines. It was enough to go to your head.

> I was called up three times but each time they sent me back to the mine.

> I was working in a shipyard in Victoria at the beginning of the war and they sent me this letter telling me to report to selected service and I didn't bother with the first one. But the next letter I got they told me exactly when to report and a week after that I was on the train coming up to Nanaimo here. When I run into this selected service I saw a guy I knew and he was the head man there. He said, "It's Cumberland for you." I said, "What do you mean, Cumberland?" He said, "Well, that's the last place you worked." "Yes, but I'm not going back to Cumberland. I'd rather join the army." So I talked 'em into letting me go to Number Ten instead.

> People that had served in the armed forces found it was an asset when they were looking for jobs at the end of the war. I was told that I was serving my country in the mine but it didn't work the same for me.

The Second World War and its aftermath gave the coal industry eight years of stability and the miners eight years of regular employment. The underground men received special rations of meat as a result of a two-week strike to protest the lack of beef in their diets, but after 1941 the government took the right to strike away for the duration of the war. That, coupled with the danger to underground safety caused by inexperienced men smoking and ignoring safety practices, started to make the mines look like they had before the union came in. The miners solved the problem of the smokers by threatening them with physical violence unless they behaved.

It had been many years since an explosion had torn apart a Vancouver Island mine. The last one had been in 1923 when Number Four mine in Cumberland went up for a second time in six months. Better safety practices and improved ventilation seemed to have eliminated that threat. Until the Beban flood it had looked as if multiple mine deaths were things of the past. Which is why it was so shocking when three men died in an explosion in Number Ten mine, South Wellington, on a Sunday morning, three days before Christmas in 1940.

The last of the Black Track mines lay at the southern end of the valley below South Wellington. In seams two and a half to six metres high, miners could make a nice wage on contract, averaging $9 for eight hours of work. Small compressed air hoists gathered the cars to the main sidings, where an electric hoist hauled coal and men up the slope to the surface. From there the loaded cars travelled a kilometre north to the tipple of Number Five mine, where they spilled their coal onto a conveyor. The coal went into a bunker and then into E & N Railway cars that took it to the wharves on the shores of Nanaimo harbour, where washers and screens prepared the coal for loading into ships at the CPR wharf.

Although the coal was mostly a low quality steam coal, the mine was modern and comfortable – a good place to work. As the workings extended farther and farther away from the main slope, however, it became necessary to construct a new intake airway. To connect the new shaft – located where the seam outcropped – with the workings, miners had been driving Number Four heading away from the main slope, carving out the necessary crosscuts as the work progressed according to the pillar and stall method.

On Saturday, December 21st, at 9:00 p.m., the crew working the weekend to earn extra money had reached the area where the heading would join the new shaft. The opening between the heading and the shaft would complete the connection but would also change the flow of air throughout the mine. The last shots of the night, however, did not appear to have broken through.

> Well on Saturday you get off an hour early maybe to go to the pub and have a beer, so they fired the last round of shots, but they broke through. Well that allowed the air to come in. Nobody worked Saturday night, but on Sunday morning they did and the new contact with the surface had short-circuited the ventilation of the mine.

Neither the evening shift fire boss nor the night fire boss had examined for gas in this area, but both of them had noted this in their records. On Sunday morning in Nanaimo, a small crew intent on earning a few extra dollars doing Sunday work scrambled aboard the jitney for South

Wellington. When the jitney pulled up in front of Jim Gava's Nicol Street house, he was running late.

> On Sunday morning Jim Gava slept in and I can still see him leaning out the window of his dark green house when the bus driver tooted the horn. He stuck his head out the window and shouted, "Go ahead, I'll bring my own car." And he got there just in time to be killed.

At 7:00 a.m., mine manager Bill Frew divided the Sunday crew into two groups. He assigned a fire boss and four men – two track layers, a hoist mechanic and a young driver – to track laying duties and another fire boss, Chris Mills, and two men – Jim Gava and Jim Waring – to the face of the new heading to clear away the rock from the previous night's shots. Mills had just acquired his fire boss ticket and at thirty-two was several years younger than the two men working under him. Whether the age difference made him reluctant to give orders or whether he neglected to heed Frew's warning to check for gas, it will never be known. When Mills, Waring, and Gava disappeared up Number Four heading at 8:00 a.m., the ventilation pattern of the mine had been short circuited for over twelve hours.

> One of them must have stayed down below to get the cars – there was cars on the rope – and when he rung the bell it must have made a spark and that ignited the gas and it killed them three men. And there was some more men down the slope straight below it and there was a fire boss with them and he said, "There is only one way we can get out. We got to put our mouth right down by the rail and crawl on our hands and knees.
>
> The other fire boss was examining for gas when this thing happened. The fire boss come back and he got a hold of us. We knew something was wrong because the explosion had sucked the air out and we couldn't breathe. The fire boss says, "Come on" and he takes us right out into the road where the explosion had gone. There was a lot of dust and we had a hard time getting out.

The track-laying crew found the rope rider, Jim Waring, lying a few feet up the heading beside two carloads of rock. Near him was the pull bell he had used instead of the regular electric bell because it might have produced a spark. But the pull bell mechanism connected two wires that completed an electrical connection and thereby defeated the whole purpose of the safer bell. Jim Waring had pulled the signal bell thinking it was safer, but it had produced a spark at the connection. A spark in a gassy environment is as dangerous as a match at the end of a dynamite fuse.

Heavy gas and smoke prevented the surviving crew from investigating further. Crawling on their hands and knees, they made their way to the surface, where they reported that the explosion had blown out a large number of timbers and caused considerable caving in one tunnel. Behind that impenetrable pile of rock and coal lay Mills and Gava.

There were several draeger teams in the area, their rescue techniques honed by hours of practice for mine rescue competitions. But this was the real thing. Of the several teams notified about the Number Ten explosion, the South Wellington team under Jock Gilmour got there first.

> We didn't have six men for the rescue team so we had to go with four. We went down the mine 1,500 feet and the air was good. But at the split we had to go under oxygen and some of the men had seen smoke on their way out. When we got down to this fire heading there it was. It was a steep uphill but right at the bottom was the rope rider, and these two cars of rock were tossed away just like two match boxes. The rope rider was thrown up against the rib just like a shirt.

> We had plenty of oxygen and my team was in good shape so we didn't wait for help but went right up the steep incline to where the hoist was. It looked like a house that had been turned upside down. Then I looked a little further I saw what I thought was a sack of rock dust until I saw a belt buckle. That was the fire boss. We kept going in and here's a big rock. The roof had fallen right down and I saw Jim Gava there. All dead. All killed just like that. They never knew what hit them. Their hair burnt white and their bodies black.

In order to reach the area where the bodies lay, the draeger crew had to restore the ventilation and repair the timbers and brattice. The process took twelve long hours during which relatives and friends of the three men gathered at the mine and re-enacted a scene that no one had thought they would see again. In Nanaimo, people crowded onto the sidewalk in front of the *Free Press* building and waited for bulletins to appear in the window. A false sense of security had been building in recent years. People had thought the mines were safer, that their ventilation systems functioned flawlessly, that the union made sure there were always thorough gas inspections.

The waiting relatives; the long vigil ending with the rescuers bringing the dead to the surface; the tears, the widows, the casting about to attach blame – it had all happened before. Nothing had changed. There would always be inexperience, greed, foolishness, always the dark mine with its black treasure and the men who would go down again.

> I used to worry my head. Oh my God. I gotta go back there tomorrow morning again. I don't know whether I'm going to make it or not. We might get out of there. We might not. Is God protecting us? Is God with us? My God, I tell you. I'll never forget it as long as I live. It was really something. Something I'll never forget.

Epilogue

The artificial boom in the coal industry inspired by the Second World War came to an end in 1947, when Imperial Oil hit pay dirt at Leduc, Alberta. In order to compete with fuel oil, coal mines had to increase their efficiency. Belts replaced shaker conveyors, portable hoists replaced mules and horses, mechanization decreased the need for labour, and jobs disappeared.

Union leaders and miners refused to believe the company reports of declining markets. They had heard it all before. But if they had been in the Canadian Collieries boardroom, they would have believed. The company was diversifying into oil drilling in Alberta and timber harvesting on its vast leases inherited from Robert Dunsmuir and Sons. The change in the company name to Canadian Collieries Resources Ltd. said it all, and that change had come several years before the company's last mine, Tsable River, produced its last coal on April 14, 1960.

Canadian Collieries Resources Ltd. made money, but not on coal. A Powell River concern bought company land; the Nanaimo waterfront buildings reverted to the CPR, BC Hydro bought the powerhouse on the Puntledge River, and Weldwood Canada Ltd. bought three hundred million metres of timber.

But the miners refused to let coal mining die. Three men took the Tsable River mine over and ran it for a few more years. They hired forty or fifty men on the basis of seniority – the youngest was fifty years old.

There were few young men interested in mining anyway. The last Nanaimo area mine, White Rapids, had closed in 1953. The younger generation had gone into logging or finished school and enrolled at university. A big pulp mill opened in Nanaimo in 1948 just when the town needed the jobs. The coal camps didn't fare so well, however. South Wellington and Extension were left to become bedroom communities for people who worked somewhere else; Cumberland reverted to village status and gave up its role as the centre of the Comox Valley to Courtenay.

Scrap metal buyers and scavengers of all sorts have spirited the pithead frames away, leaving only the concrete foundations and motor mounts hiding in the lush Vancouver Island vegetation. Water rushes out of hidden tunnels; the ground subsides into abandoned workings; black slack heaps nurture mature trees; pilings rot on harbour shores. But the miners remain alive in their words.

So I hope that's going to interest the younger generation. I hope it's going to open their eyes and ears. What I said here is the honest truth. I've had to struggle. We all had to struggle. We struggled worse than any human being. But us coal miners we really made up our mind that we were going to do our best, union or no union. And we kept our word and we gained in the long run.

I worked thirty-three and one half years for this company and when I folded up I was still $100 in the hole.

We had our good times too. Some places I had a good job and other times I was pretty well heavy worked. I had some good mules and I had some tough ones. I liked diggin' coal. If you had a pretty good place with a fairly good amount of coal, a good roof and a good floor, it was pleasant work. We had our good times.

What time did I like best of all? Quittin' time.

Notes

Abbreviations:

CTHP	Coal Tyee History Project Tapes and Transcripts – deposited at Malaspina University-College and the British Columbia Archives
NCA	Nanaimo Community Archives
RCID	Royal Commission on Industrial Disputes in the Province of British Columbia, 1903

Chapter I – Boss Whistle

"I first went down as a winch boy…" Wilf Brodrick, CTHP 4051:16

"One place I heard this rumble…" Alex McLellan, CTHP 4051:86

"You couldn't see. You had to go…" George Mitchell, CTHP 4051:91

"I never got lost underground…" Alex (Mike) Raines, CTHP 4051:103

"I think it's against nature…" Dominic Armanasco, CTHP:4

"It's dangerous sometimes but…" Glyn Lewis, CTHP 4051:72

"The only thing I was afraid of…" Mima Sheppard, CTHP 4051:108

"I never thought of the mine…" John Irvine, CTHP 4051:63

"My mother mighta worried…" Henry Gueulette, CTHP 4051:56

"In 1909 there was an explosion…" Jock Gilmour, CTHP 4051:49

"My father came out in 1910…" Laura and Tom Johnston, CTHP 4051:65

"When they're down in the mine…" John Pecnik, CTHP:98

"Even if they're not friendly…" John Gourlay, CTHP 4051:52

"We earned every nickel we got…" George Edwards, CTHP 4051:39

"In Granby there was a great number…" Jack Unsworth, CTHP 4051:22

Chapter II – The Put-Up and Pull-Down-Again Towns

"Oh Extension is the Devil's…" Mrs. Frank Wall, CTHP 4051:124

"It was the end of the world…" Albert Steele, CTHP 4051:113

The origins of Extension and Ladysmith: Louis Astori, Jonathan Bramley, John Bryden, James Dunsmuir, Joseph Fontana, George Johnson, William Joseph, James Pritchard, Joseph Tassin, Moses Woodburn testimony, RCID

"'Take your houses with you…'" Waino Torkko, CTHP 4051:121

"It was cheaper…" George Johnson testimony, RCID

"And you'd go up a hill…" Wall, 124

"Extension was just like…" Torkko

"And the roads in Extension…" Wall, 124

"What was Extension like…" Dave McDonald, CTHP 4051:81

"When Dunsmuir first decided…" Eino Kotilla, CTHP 4051:67

Dunsmuir's decision to move the wharves: Bowen, Lynne, *Three Dollar Dreams*, pp. 359-60

Extension school: Mark Fontana interview, NCA

"If my mother wanted anything…" George Mrus, CHP 4051:93

"My mother was only here…" George Bodovinick, CTHP 4051:14

"I went up to the mine…" Mrus

"I would be about…" Torkko

"I've been in the graveyard…" Mitchell

"There was lots of noise…" Steve Arman, CTHP 4051:3

"The rock dump…" Elmer Blackstaff, CTHP 4051:13

"Everyone called that rock dump…" Kotilla and Arman

"They had a room…" Arman

I serviced the lamps…" Ibid.

There were no lights…" McLellan

I think there used to be…" Kotilla and Mitchell

When you're in the mine…" Armanasco

"Course Extension was high seam…" Mitchell and Arman

"Now this is what I mean…" Jack Atkinson, CTHP 4051:5

"It's hard to explain…" Gilmour, 49

"You'd try and chase them…" Torkko and Bodovinick

"Oh there's some mean mules…" Bodovinick

"I drove this big Blacko…" Mitchell

"I drove one by the name…" Ibid.

"Oh yes, they'll drive you…" Gilmour, 49

"In some cases the roof…" Armanasco

"There was a hundred…" Dave McDonald

"There was about two feet..." Bob Winthrope, CTHP 4051:128

"I've worked in the wet..." Armanasco

"When the cars got outside..." Tom McDonald, CTHP 4051:82

"There was a lot of Chinese..." Ibid.

"I was only a school boy..." Edwards

"In Extension mine..." Ibid.

"Extension mine wasn't much on safety..." Tom McDonald

"Well when I was a kid..." Pecnik, 98

"My dad lived through..." Bodovinick

"They had a good bucket..." Arman

"Now the Bowaters came..." Archie Greenwell, CTHP 4051:53

"We were like one family..." Armanasco

"They had a dance here..." Arman

"We put in about seven..." Ibid.

"We didn't make any money..." Unsworth

"I was workin' with a bunch..." Jock Craig, CTHP 4051:31

"After the Extension mines closed..." Arman

"As long as I've been around..." Hjalmar Bergren, CTHP 4051:8

"The fact that they were so far..." Unsworth

"Extension was kind of a rough..." Ibid.

Chapter III – The Black Track Mines

"My grandfather, Sam Fiddick..." Dick Fiddick, CTHP 4051:42

"There were prospectors..." Ibid.

"If you took the land..." Lewis Thatcher, CTHP 4051:117

"Now the way I understand..." Myrtle Bergren, on Fiddick tape

"It was gonna cost a little..." Ibid.

"They had to go to court..." Thatcher

"Well the outcome of it was..." Fiddick

"Fiddicks was sellin'..." Thatcher

"Mrs. Fiddick was a good..." Steele

"Timber 130, 140 feet high..." Thatcher

"It took me half a day..." Steele

"I can remember seeing the oil lamps..." Gourlay

"We used to use dogfish..." Edwards

"My mother lived on that little island..." Victor Balatti, CTHP 4051:6

"Many a times there were little holes..." Edwards

"I've got an old safety lamp..." Pecnik, 99

"At first everybody had these..." Ibid.

"First thing, you were searched..." Ibid.

After 1917, we had electric..." Torkko

"Well there's no pressure..." Steele

Well, just how shall I say it?..." Atkinson

"William Gibson, who survived..." *The Daily Colonist*, March 30, 1975, p.6

"There was another case..." Atkinson

"I had to go and identify..." Steele

"They found a guy sittin'..." Craig

"Most of the fellows..." Steele

"It was greed you see..." Atkinson and Gilmour, conversation with author

"They should have known..." Steele

"The miners' union while still organized..." *Free Press*, February 9, 1915

"There was nothing for the widows..." Atkinson

"My dad was killed in 1922..." Archie Greenwell

"It was the house next door..." Ibid.

"Well that was across the lake..." Thatcher

"In the early days..." Ibid.

"There were some Italians..." Ibid.

"Now the Finns came out..." Ibid.

"There was Finns and Italians..." Gourlay

"They came from the Old Country..." Ibid.

"It didn't seem to matter..." Ibid.

"By golly there was always..." Craig

"We had three trains each day..." Pecnik, 98

"The bus went to Nanaimo..." Marvin Thomas, CTHP 4051:118

"In the summer time..." Archie Greenwell

"We used to come home..." William Crawshaw, CTHP 4051:32

"Well when you were a boy..." John Sandland, CTHP 4051:107

"They couldn't keep me out..." Ike Aitken, CTHP 4051:2

"The only reason I went down..." John Carruthers, CTHP 4051:22

"I never had my own..." Kotilla

"The thing I remember most..." Tom Dixon, CTHP 4051:38

"I started picking rock..." Atkinson

"The older men pulled little..." Atkinson and Gilmour in conversation with author

"We were on a place..." Malcolm Odgers, CTHP 4051:95

"It was a horse I had the worst..." Gilmour

"Sometimes I used to get..." Torkko

"A mule was worth more..." Ibid.

"Always called a rat..." Carruthers, 23 and Pecnik, 98

"And when the pumps got all..." Torkko

"The Ida Clara..." Dolly Gregory, CTHP 4051:55

"My mother used to say to me..." Ibid.

"One of her cows got killed..." Thatcher

"I went back to Southfield..." Pecnik, 99

Chapter IV – The Town That Fought Back

For more detail on the first decade of Cumberland history see Bowen, *Dreams*

"In the old days..." Mima Sheppard, CTHP 4051:108

"I sold beer in the King George..." Alex Dean, CTHP 4051:35

"The one you had to watch for..." Johnny Robertson, CTHP 4051:106

"In them days there were..." Jim Galloway, CTHP 4051:46

"The miners' entertainment?..." Alex Dean

For more detail on anti-Asian legislation and the CMR Act see Bowen, *Dreams.*

"The white men hired Chinese..." Sam Cameron, CTHP 4051:20

"The old timers would take..." Steele

"There's not that many Chinese women..." Ed Lee, CTHP 4051:70

"That's when they dress up..." Sheppard and Rhoda Beck, CTHP 4051:7

"Us white kids always loved..." Ann Bryant, CTHP 4051:17

"On the last day of the year..." Lee

"My brother and mother used to get mad..." Sheppard

"They had those silver opium..." Gilmour, 49

"They were good workers...." Sam Cameron

"I worked in Cumberland Number Four..." Gilmour, 49

"You didn't really have any trouble..." Steele

"If they laid off one Chinaman..." "Wink" English, CTHP 4051:41

"I worked with the Chinamen..." Ibid.

"The 'other' race..." Gourlay

"We were scared I'm telling you..." Ibid.

"They were the most friendliest..." Sheppard

"When a new seam started..." Sam Cameron

"The government that stopped..." Lee

"I can remember the Chinese..." Janet Robertson, CTHP 4051:106

"If you tried to ask..." Steele

"I think the Japanese were more..." Abbondio Franchesini, CTHP 4051:43

"At the time the war broke out..." Harry Ellis, CTHP 4051:40

"When my mother and my two..." Janet Robertson

"People weren't prejudiced..." Sheppard

"Some people blame the Chinese..." Sam Cameron

"My father told me it was so bad..." Ben Horbury, CTHP 4051:60

"The mines I seen the most gas..." Torkko

"Number Five was dangerous..." Sam Cameron on Margaret Cameron tape, CTHP 4051:19

"Except for one level..." Johnny Robertson

"I was a haulage man..." Ibid.

"One of the biggest troubles..." John Peffers, CTHP 4051:100

"You get used to working..." Alex Dean

"There's two advantage to workin'..." Jack Atkinson conversation with author

"You wouldn't catch me workin'..." Gilmour conversation with author

"Longwall is like working under..." Syd Tickle, CTHP 4051:120

"Two or three guys got hurt..." Johnny Robertson

"Pretty hard to describe..." Peffers

"I was one of the diggers..." Ibid.

"This shaker shakes the coal..." Horbury

"They took out enough rock..." Peffers

"I've seen one. It's really scarey..." Pecnik, 98

"Funny, in the mines..." Ibid.

"Aw, baloney! If it's gonna..." Gilmour conversation with author

"I can't tell you any funny stories..." Unknown

"A lot of people in Nanaimo..." Sam Cameron on Margaret Cameron tape

"I was up to my waist..." Sam Cameron

"My father was killed..." Margaret Cameron

"There were thirteen children..." Ibid.

"Oh they were smart. A white kid..." Isabel Merrifield in Judith Hagens' *Chinatown, Cumberland*, p. 27

"Nobody seems to know why..." Alex Dean.

"These were all company houses..." Mabel Williams, CTHP 4051:125

"When we first came we didn't..." Ibid.

"At the peak there were..." Ibid.

“Number Eight where I was workin’…” Alex Dean and Arman

“They had a great big fan…” Mabel Williams 125

“I guess if something happened…” Ibid.

“There was two men killed…” Alex Dean

“They took everything out…” Mabel Williams, 125

“When the mine closed down…” Ibid.

Chapter V – In the Shadow of Number One

For a more comprehensive discussion of the 1887 explosion see Bowen, *Dreams*, Chapter VII

“It was five minutes…” Elizabeth Freeman, CTHP 4051:44

“When my mother was a girl…” Beck

“Some of the victims…” Alan Hall, CTHP 4051:57

“There used to be a fence…” Bill Cottle, CTHP 4051:30

“When old Robert Dunsmuir…” Ibid.

“The old Southfield mine…” Beck

“Everyone was afraid of Fox…” Pecnik, 98

“So maybe we’re coming off shift…” Aitken

“The cars generally weighed…” Tom Terry, CTHP 4051:116

“If you didn’t come back out…” Crawshaw

“My first job was runnin’ winch…” Edwards

“There was a man who used to ride…” Cottle, 30

“Canadian Collieries and Western Fuel…” Lewis, 72

“Von Alvensleben. Yes, he had this Jingle Pot mine…” Cottle, 30

“There was a German fellow here…” Ibid.

Von Alvensleben and Auchinvole: In the first edition of *Boss Whistle* the author did not make it clear that the miners’ information about these two men was inaccurate. The name “Auchinvole” is a Scottish name not a German one. The family originated in Kilsyth, Ayrshire. Two brothers, Alexander and Henry, emigrated to Vancouver Island. Alexander worked for the Dunsmuirs in Wellington, Extension and was Superintendent of Canadian Collieries at Union Bay. Henry worked in the office at Union Bay but his great-niece, Joan Auchinvole Hurn, knows little else about him. It is conceivable that he had some association with the Jingle Pot mine but he was not involved in Von Alvensleben’s nefarious activities. The author is indebted to Graham Holland for bringing this to her attention, to Mrs. Hurn for supplying more details, and to the late David Ricardo Williams’ book, *Call in Pinkerton’s*, for more information on Von Alvensleben.

“The timekeeper lived in an old…” Lewis, 72

“One morning it blew…” Nelson Dean, CTHP 4051:36 and Edwards

“You had to be real careful…” Edwards

“One Sunday some carpenters…” Ibid.

“We weren’t afraid…” Ibid.

“Oh, I guess we figured it was nicer…” Albert Tickle, CTHP 4051:119

“They must have had quite a bit of stuff…” Cottle, 28

“There was ten Chinamen…” Edwards

“They said there must have been over 800 tons…” Balatti

“Those explosions always seemed to take place…” Edwards

“I was blendin’ for the gelignite…” Balatti

“We lived on Kennedy Street…” Nelson Dean

“We had a house up on Victoria Road…” Beck

“When that boat exploded…” Mrs. John Hunt, CTHP 4051:62

“We was all workin’ at the powder works…” Cottle, 28

“Now this boat, they called the S.S. *Oscar*…” Ibid.

“Well in their excitement the crew lost…” Ibid.

“Somehow or other this sulphur…” Ibid.

“As soon as the plant moved…” Ibid.

“We worked in about a three-foot…” Blackstaff, 13

“I didn’t like working the low seam…” Joe Krall, CTHP 4051:68

“The first shift I went on in Protection…” Joseph Melissa, CTHP 4051:88

“They couldn’t work it because it used to take fire…” Peffers

“The *Patricia* was the passenger boat…” Edwards

“You see, the war was on at the time...” Nelson Dean

“The rope broke. I guess it couldn’t…” Edwards

“I saw the sixteen miners from Protection cage…” Syd Tickle

“We paid two fellas two bits…” Scotty Gilchrist, CTHP 4051:47

“The lamp man would say…” Gilmour, 48

"In a mine nothing can stop the coal..." Ibid.

"The regulations were good..." Ibid.

"And we took the training..." Ibid.

"If there was a mine explosion..." Alex Dean

"My partner was my brother-in-law..." Peffers

"My father got hit by a runaway..." Harry Dawes, CTHP 4051:33

"It was very important to a miner's wife..." Syd Tickle

"You always tried to have dinner..." Myrtle Bergren on Armanasco tape

"When he leaves the house..." Anonymous, *Vancouver Morning Sun*, February 12, 1913

"My father and another man..." Odgers

"There was going to be a great time..." Mike Plecas, CTHP 4051:101

"There were narrow spots and then..." Pecnik

"So I'm runnin' head down..." Ibid.

"We went up so fast..." Ibid.

"First of all there's the scow..." *We Too*

"Of all the appalling conditions..." Ibid.

"That mine rock caught fire..." Blackstaff, 13

"In the olden days you very seldom..." Lee

"Lots of people came here..." Ibid.

"Instead of cook for themselves..." Ibid.

"One Chinese was the leader..." Gourlay

"The 'bossy man' ran a real racket." Atkinson and Gilmour conversation with author.

"The contractor was 'above' coal mining..." Hall

"On Saturday night the population..." Lee

"I looked after the motor..." Carruthers, 22

Chapter VI – The Union Makes a Stand

The Dunsmuirs had sold their railway holdings to the CPR in 1905.

"I object to all unions..." James Dunsmuir, RCID

Enterprise Union: The author is indebted to Kathy and Garth Gilroy for allowing her to read the minute book of the Enterprise Union, which has been in the possession of the Gilroy family since the early years of the twentieth century.

"I asked them where the nigger was..." James Dunsmuir, RCID

The Socialist Party of British Columbia (SPBC): Jean Barman in her *The West Beyond the West*, credits the SPBC with originating socialism in Canada in 1901. A national party emerged three years later. According to Dorothy Steeves in *The Compassionate Rebel*, the history of socialist and labour parties in B.C. is a complicated one full of factions and feuds. In 1909 the *Western Clarion* said, "Oil and water would mix as readily as unionism and socialism." But during the big strike, the attitude of old socialists toward unions softened somewhat.

Eight hours bank to bank: the length of the miners' working day could vary considerably depending on whether their work day was calculated from when they left the pithead (the portal) or from when they reached the face (bank).

Natal Act: According to Steeves, who is quoting the *B.C. Federationist*, Dunsmuir refused to give royal assent to the act, which required an education test for immigrants and would have excluded Chinese. Hawthornthwaite charged that at the time Dunsmuir had a contract to "procure 500 Japanese coolies for exploitation in his coal mines."

"You had to produce otherwise..." Gilmour and Atkinson conversation with author.

"Because if you haven't..." Dawes

"A human being was nothing." Ibid.

Deaths from gas explosions: the number 373 comes from the *B.C. Federationist*, September 7, 1917. The number is accurate according to the *Ministry of Mines Reports* for those years.

"They had gas committees..." Robertson

"If you smelled gas..." Dawes

"If you reported gas..." Ibid.

"Who wouldn't fight..." Anonymous, *Vancouver Morning Sun*, February 12, 1913.

The situation in Cumberland before the strike: See John Norris, "The Vancouver Island Coal Miners, 1912-1914: A Study of an Organizational Strike," *BC Studies*, no. 45, Spring 1980 and Alan John Wargo, "The Great Coal Strike. The Vancouver Island Miners' Strike, 1912-1914." Unpublished B.A. Graduation essay, University of British Columbia, 1962

"reserved the right..." Wargo.

"This is what was supposed..." Atkinson

"But I think the company..." Galloway

"Up in Cumberland you see..." Ibid.

"Yes, you couldn't blame..." Ibid.

"I heard about one Italian fellow..." Horbury

"The name scab must have come..." Robert Ross Loudon, CTHP 4051:77

"We called 'em black Minorcas." Edwards

"When we were kids..." Margaret Cameron

"My father worked nine shifts..." Dawes 33

"No matter what they do after…" Ibid.

"When the strike began…" Atkinson

Strikebreakers: Both Wargo and Norris agree that many of the strikebreakers were former strikers although Wargo's estimate of the numbers was higher than Norris'. Using figures from the *B.C. Federationist*, Norris reports the following: In February 1913, eighty-nine strikebreakers were recruited in Edmonton but only fourteen reached Cumberland; in June 1913, sixty-nine were recruited in Durham and forty of these were working in Cumberland in August 1913; in the same month twenty-three Italian labourers came from the United States to work in Extension; in August 1913 forty-three Italian and British men from Extension, eleven of whom were refugees from the Extension riots, were recruited for Cumberland but only thirty-seven actually reported for work. The PCC recruited a number of strikebreakers in Joplin, Missouri.

"One of my sisters was born…" Dawes

"So we came to Cumberland…" Lewis, 72

"We moved to Bevan in 1911…" Jean Cameron Wheeler on Margaret Cameron

"I talked for forty-five minutes…" Parker Williams quoted in the *Vancouver Morning Sun*, February 12, 1913

"You don't forget when you see…" Ibid.

"My mother had things tough…" Gilmour, 48

"So it kind of come hard times…" Cottle, 28

"I was two years on miners relief…" Galloway

"A man and his wife got six dollars…" Matthew Clue, CTHP 4051: 26

"I don't know how we did it…" Bodovinick

"We got put out of the house…" Dawes

"We lived on mush and beans…" Sam Cameron

"As I recall, during the strike…" Archie Greenwell

"Mother backed Dad up to the limit…" Gilmour, 48

"We all had gardens them days…" Craig

"I went through the strike…" Johnston, 66

"In fact, many a young man…" Lempi Guthrie's account written for Myrtle Bergren and used with Guthrie's permission

"There was lots of things you had to go without…" Galloway

"There was a woman whose husband…" Atkinson

UMWA takes the strike to the whole island: See Norris; Frank Farrington to Robert Foster, April 30, 1913, Parker Williams Fonds, NCA.

"They went over to England…" Jim Galloway in *Fighting for Labour: Four Decades of Work in British Columbia, 1910-1950*, Derek Reimer, ed. *Sound Heritage*, Volume VII, Number 4, 1978

"They were told: 'Don't go to Cumberland…' Ben Horbury, *Ibid.*

"The women was with the men…" Galloway

"The women did a lot…" Gilmour, 48

"We had a seven-piece…" Unknown

"One of the special policemen…" Atkinson conversation with author

"The specials belonged to…" Anonymous, *Free Press*, June 28, 1913

"The guys who couldn't stick it out…" Irvine

"In them days the papers…" Galloway

"Not until it hears with its own ears…" *Free Press*, May 30, 1913

Chapter VII – Hurray, Hurray, We'll Drive the Scabs Away

"The law in Ladysmith today…" Parker Williams

"The fight has not been a spectacular one…" *Nanaimo Free Press*, May 31, 1913

"In August of 1913…" Edwards

"I was seven years old…" Kotilla

"You could hardly hear your own ear…" James White, testifying at E. Morris preliminary hearing, September 2, 1913

"They had a heck of a splatter." Craig

"The Ladysmith men are a distinctly lower order…" David Ricardo Williams, *Pinkerton's*

"A whole bunch of young people…" Atkinson

"There was men and horses…" Ben Horbury in *Fighting for Labour*

"I was on the street…" Horbury

"I remember the riot up there…" Jean Cameron Wheeler on Margaret Cameron tape

"So the story got down here…" Atkinson

"The company got all the bums…" Gilmour, 48

"My father came up from…" Tom McDonald

"A lot of Italians came in…" Mitchell

UMWA role in starting the riot: David Ricardo Williams's *Pinkerton's* notes the presence of Farrington, Walker, Irvine and Angelo. Norris provides convincing evidence that outside union leaders provoked the riots and ran the strike. Wargo confirms that Joe Angelo was the Italian organizer and held an important position in the international union but he blames "a significant radical element [among the strikers] which was subject to rash and thoughtless action" for the violence. The *B.C. Federationist* is full of "radical socialist" rhetoric as Wargo calls it. David Ricardo Williams's Pinkerton detective had no use for "radical socialists" but conceded that "less radical socialists" like Parker Williams were entitled to respect.

Italian immigrants: See Bowen, *Dreams*

"We were at Number One…" Ellis

"My dad had a little…" Atkinson

"Around midnight my dad…" Ibid.

"A friend of mine was farming…" Ben Richardson testifying at the Louis Nuenthal preliminary hearing, September 9, 1913

"A bunch of strikers found…" Ibid.

"During the day, a number…" David Williams in his "Memoirs of the Vancouver Island Coal Strike" by permission of his nephew, Parker Williams, Jr.

"The pedestrians laughed at him…" Lempi Guthrie written account

According to Steeves, "Hurray, hurray, we'll drive the scabs away" may have been a variation of a song written by one of the strikers, Bob Millis, who composed the music and words to that other strike anthem, "Bower's Seventy-Twa." Philip J. Thomas in his *Songs of the Pacific Northwest* credits a *Bill Willis* as one of the composers and adds that Willis was arrested for singing it. According to the miners in the Coal Tyee Society, Bill and his father, Joe Willis, made the song up and sang it at all union meetings.

"She picked up an axe…" Ibid.

"In the course of the evening…" David Williams, "Memoirs"

"Shortly after twelve o'clock…" Thomas O'Connell, *Ladysmith Chronicle*, August 23, 1913

"There was a man in Ladysmith…" Thatcher

"Mr. McKinnon had no strong…" Guthrie

"There were two fellows who…" Ibid.

"McKinnon heard a noise…" Thatcher

"This incident reflected badly…" Guthrie

"With regard to Stackhouse and…" Ibid.

"Feeling was running very strong…" Ibid.

"There was a strong rumour…" Atkinson and Gilmour conversation with author

"By now it was August 13…" David Williams, "Memoirs"

"I heard the two explosions…" Robert Bickerton, *Ladysmith Chronicle*, October 14, 1913

"I can remember as a little kid…" Gourlay

"Six o'clock the next morning…" Guthrie

"A burly Scotch miner…" Ibid.

"One policeman asked Guthrie…" Constable A.C. Pallant testifying at the "Speedy Trial" from *Ladysmith Chronicle*, October 14, 1913.

"Later my dad received…" Gourlay

"All those visits did was scare…" Anonymous, *Ladysmith Chronicle*, August 23, 1913

"I guess this had been planned…" Kotilla and Gourlay

"I talked to one of the Ladysmith rioters…" Gourlay

"Sam was very opposed…" Guthrie

"Sam Guthrie was one of the leading…" Gourlay

"Sam Guthrie used to come over…" Kotilla

"Sam Guthrie was a good man…" George Bryce, CTHP 4051:18

"At 11:00 a.m., a number of men…" James Roaf, Nuenthal hearing, September 9, 1913

"They had a bullpen…" Gilmour, 48

"Well you know drunken men…" Ibid.

"We were staying in the Cinnabar…" Tom McDonald

"At that time we were…" Ellis

"Well, if you've done anything…" Gourlay

"So I got a bunch of guys…" Ellis

"I can remember seeing the strikers…" Percival John testifying at W. Bowater trial, January 12, 1913

"The people from Nanaimo seemed…" George Hannay testifying at F. Morris preliminary hearing, September 2, 1913

"The crowd was coming down…" Charlotte Schivardi testifying at Morris hearing

"The union gathered that day…" Ellen Greenwell in notes accompanying Archie Greenwell transcript

"They all had guns and they all…" Gilmour, 48

"Fifty or sixty men…" Percival John at Morris hearing

"Buckshot was coming right out…" Frank Beban at Morris hearing

"My grandparents lived in Extension…" Gourlay

"The union men tried to get…" Thatcher

"There was two or three hundred people…" Percival John at Bowater trial

"It was war..." Edwards

"Alec Baxter was up there..." Thatcher

"To my knowledge I don't think..." Archie Greenwell

"And there was one fellow..." Gilmour, 48

"So, anyhow, when a bunch..." Ellis

"The fellows came up from Nanaimo..." Gilmour, 48

"I saw a large crowd of men..." Wong Back Chung testifying at Investigation of Damage to Chinese Property, August 19 and 20, 1913

"The first we saw was the end..." Percival John at Morris hearing

"I heard shooting between..." James White at Morris hearing

"I saw a man playin'..."

"They told us their was no more Extension..." Thomas Isherwood to Emily Isherwood, August 15, 1913

NINE MEN SHOT IN EXTENSION TODAY: In the subsequent investigation into this incident, the *Free Press* denied responsibility. It was common practice for newspapers of the day to post major news stories in their front windows so that passersby could get the news before the papers hit the streets. The *Free Press* said no such notice was placed in its window on Wednesday, August 13th. If this is correct, then the actions prompted by the rumour were strictly the result of what people heard on the street. There is no doubt that the *Free Press* published the headline later that day. And a subsequent edition told of six strikers being shot by scabs, ten men being wounded and the exodus of the doctor with supplies and men from Nanaimo on horseback.

"I was near Quennell's..." Wong Fung testifying at Investigation of Damage to Chinese Property

"My dad was a fire boss..." Jock Ovington, CTHP 4051:97

"Them shingles was comin' down..." Dave McDonald

"A great deal of plunder..." Atkinson

"Cunningham, the manager, had..." Gilmour, 48

"I was fourteen years old..." Blythe Cosier testifying at Bowater trial

"I owned a store in Extension..." Wong Fung at Investigation into Damage to Chinese Property

"During the strike there..." Lorenzo Giovando, CTHP 4051:50

CHAPTER VIII – BOWSER'S SEVENTY-TWA

"Then they brought in the militia..." English

"And the people in Nanaimo said..." Ellen Greenwell in *Fighting for Labour*

"They'd been workin' on the road..." Anonymous

"Well all that trouble led..." Ellis

Hamber and Currie: Some of the officers posted to Vancouver Island after the riots would become famous. Eric Hamber beame the lieutenant governor of B.C. from 1936 to 1941. Arthur Currie was the first Canadian to command the Canadian Expeditionary Force and was knighted for his role in the capture of Vimy Ridge during the First World War.

"Remember all them soldiers..." Balatti

"The militia closed the bars..." Giovando, 50

"I was a kid goin' to school..." Torkko

"Col. Hall said..." Ellis

"Well I can tell you..." Ellen Greenwell notes

"...a true blood both union-wise and political-wise" Archie Greenwell in an interview by Barbara Simkins in Lake Cowichan, B.C., August 30, 1989

"So by this time..." Ibid.

"My father owed the CPR for the fare..." Gilmour, 48

"...were positive they saw rifles..." Captain W. Rae in his "Narrative and Appreciation of Events on Vancouver Island, August 14th-31st, 1913" in Reginald H. Roy's "The Seaforths and the Strikers: Nanaimo, August 1913" *BC Studies*, No. 43, Autumn 1979

"My father was clearing land..." Cottle, 28

"My father was the secretary..." Bryant

Place and Williams: A supporter wrote from England to Williams in November 1913, "McBride and his gang would have full swing if you two were off the role. They would have no one to administer justice to them." G. Arthur Bagshaw to Parker Williams, November 17, 1913, NCA

"I was on a load of oats..." David Williams, "Memoirs"

According to information supplied by Pam Beardsley, her great-grandfather, Joe Taylor, was arrested by detectives in Duncan on August 17th on his way to a meeting.

"I note that the wives..." David Ricardo Williams, *Pinkerton's*

"Seldom does it fall..." *Ibid.*

"So we had to visit my father..." Bryant

"At first visitors were allowed..." David Williams, "Memoirs"

"When my father was in jail in Nanaimo..." Kotilla

"Mother couldn't go to see Dad..." John Slogar, CTHP 4051:111

"My father-in-law didn't say..." Bryce

"No, but I did join in..." Guthrie

"She simply turned the tables..." Ibid.

"The outcome of the trial..." Ibid.

Picketing: The decision was rendered by the Judicial Committee of the Privy Council which was at that time the court of final appeal for Canada.

"It is a positive fact..." David Ricardo Williams, *Pinkerton's*

[T]heir skirts go up so far..." *Ibid.*

"Men can't fight against guns..." Nelson Dean

"We got off the train..." Henry Wilby letter to author

"Your counsel has said all..." Frederick W. Howay in the *Ladysmith Chronicle*, October 14, 1913

"I have looked over your faces..." *Ibid.*

"There were six of us kids..." Slogar

"And I remember when they went..." Gilmour, 48

"I ran across the Judge..." Israel I. Rubinowitz letter to Parker Williams, February 1, 1914, NCA

"Uncle Bill Bowater..." Donald (Dusty) Greenwell, CTHP 4051:54

"If McBride or any of his gang..." Robert Gosden quoted in Steeves, *Compassionate*

"The new miners now being employed..." David Ricardo Williams, *Pinkerton's*

"Some had worked all along..." Tom McDonald

"My dad was a fire boss..." Ernie Johnson, CTHP 4051:64

"I guess a lot of people were beginning..." Ovington

"I finally asked my father about it..." Nelson Dean

"Vancouver Island coal mines will be organized..." *Free Press*, May 1, 1914

"...lickspittles of strikebreakers..." *Free Press*, July 14, 1914

"I remember when they got out..." Gilmour, 48

"From then on it was kind of chaos..." Ibid.

"Well certainly it was..." Ellis

"I don't think there's anybody..." Thatcher

"Well, apparently in South Wellington..." Tom McDonald

"It was always suspected..." Steele

"The South Wellington fire..." Craig

"My dad was on his way back to work..." Cottle, 28

"If you were blacklisted from one mine..." Dawes

"During the strike there was a lot..." Bodovinick

"It was a very, very bad thing..." Irvine

"My father was unemployed..." Ibid.

"My father was blackballed..." Slogar

"After the big strike, father was kep'..." Aitken

"There was a mine near Lethbridge..." Mrus

"So eventually Dad went up to Fernie..." Cottle, 30

"There was no welfare..." Ibid.

"That's right. That's how I got out..." Steele

"After the South Wellington flood..." Gilmour, 48

"My mother was bathing us kids..." Atkinson conversation with author

"I felt sorry for the scabbies..." English

"All the scabs that was livin'..." Craig

"There was enough scabs in Cumberland..." Dawes

"There was a lot of bitterness..." Arman

"There was hate in that strike..." Gilmour, 48

"There were families split down the middle..." Archie Greenwell

"Well I know one family..." Dusty Greenwell

"Some people were very religious..." Tom McDonald

"It was sad because the Plymouth Brethren..." Ibid.

"Some families had miners and farmers..." Archie Greenwell

"After the strike my father..." Ibid.

Chapter IX – Haulin' Coal and Arguin'

"My family used to camp..." Beck

"I was twenty years old..." Alex Menzies from notes with CTHP:89 transcript

"It wasn't such a great mine..." Albert Tickle

Suquash: The location of the first coal mined in British Columbia in 1835 when First Nations people picked the surface coal. The HBC mined coal briefly there around 1850. PCC reopened it in 1908 but it was idle for most of the First World War.

"Well I stayed clear..." Pecnik, 88

"A rock miner who was driving..." Odgers

"Nobody worried about coal dust..." Clarence Hamilton, CTHP 4051:59 and Hall

"We were working the rock tunnels..." Lewis, 72

"I remember in Reserve..." McLellan

"You could smell sulfur..." Gilchrist

"You worked one day..." Crawshaw

"They phoned me one morning..." Hall

"Something had broken on the main..." William Raines, CTHP 4051:104

"One of the fire bosses..." Dave McDonald

"We had our big shower..." William Raines

"Now a blowout is different..." Unsworth, 122

"It always gave a warning..." Steele

"And the fire boss brings..." Mitchell

"My brother and I were working..." Ibid.

"It was a dangerous mine..." Steele

"I was going to a miners' meeting..." Dave McDonald

"We both lived in South Wellington..." Amy Craig on Jock Craig tape

"I didn't get married until..." Balatti

"There was an acceptance of characters..." Johnston, 66

"At first the miners used to get..." Nelson Dean

"The only thing I haven't forgotten..." Giovando, 51

"When the flu epidemic..." Sheppard

"During the big flu in 1918..." Irvine

"A bunch of us got together..." Slogar

Payday: Weir Muir, "Fond Memories of Victoria Road," *Nanaimo Daily News*, May 8, 2000

"The wages were very poor..." Gilmour, 49

"I worked Timberland mine..." Dawes

"They had a commission here..." Gilmour, 49

"The beer parlour..." Balatti

"You don't know how much..." Hall

"When I was a little chap..." Gourlay

"When I was driving in Number One..." Mitchell

"There were fifteen hotels..." Ibid.

"We always went to town on Saturday..." Balatti

"Saturday was the day..." Jack Ostle, CTHP 4051:96

"Well I don't remember the grown-up..." Bryant

"I didn't know what they meant..." Beck

"Before I went in the mines..." Ostle

"The only time I was on Fraser..." Ernie Johnson

"Miners went there mostly..." Krall interview notes

"It was mostly the sailors..." Sandland

"Just for big shots..." Winthrope

"If it wasn't for the mine men..." Arthur Simpson, CTHP 4051:110

"My people were like most..." Bodovinick

"They didn't go much on religion..." Herschel Biggs, CTHP 4051:10

"I was supposed to go to church..." Dave McDonald

"We'd speak on street corners..." Johnston, 65

"My mother and father were..." Dave McDonald

"When Mother and Father arrived..." Hugh Campbell-Brown, CTHP 4051:21

"Oh, I was fond o' dancin'..." Edwards

"And you could fire a shot..." Gilchrist

"Skipper" Moore: Interview with Moore's daughter, Anne Clark

"If you played football..." Robertson

"Most football players were Scotch..." Armanasco

"If you were a good soccer..." Ernie Johnson

"They always talked football..." Ibid.

"Football and baseball were the big things..." Horbury

"You had to work there so long..." Sandland

"They'd ask you all these different..." Mike Plecas

"My dad used to say your basic..." Ronald Williams on Mabel Williams, 126

"A miner himself has got to be..." Dusty Greenwell

"I passed the exam and got..." Alex Dean

"My father says..." Lewis, 72

"That's not very deep..." Mrus

"They sunk a slope for getting air..." Ibid.

"They used to ship the coal..." Ibid.

"They gave 'em permission..." Mrus and Dawes

"The ground hasn't changed..." Mrus

"It had a bench on each side..." Ernie Johnson

"They'd be waitin' there at the mine..." Odgers

Chapter X – We Too

"Even an animal will fight..." Gilmour, 48

"In the old country the union..." Bryce

"If the word got out that you were..." Ostle

Constable Campbell: Police charged Campbell with manslaughter, but a grand jury meeting *in camera* exonerated him. Ginger Goodwin's defenders have since pointed out many discrepancies in the official record.

"I joined in 1922..." Gilmour conversation with author

"There was a union tried to come in..." English

"My dad had an OBU button..." Dawes

"Before we had the union..." William Raines

"One of the head men..." Carruthers, 23

"Before the unions come here..." Edwards

"It wasn't the fourteen-inch seams..." Carruthers, 23

"When I started in the coal mine..." Ibid.

"Before you were a slave..." Gilmour, 48

"Fred and I worked together..." Carruthers, 23

"Well we were treated like rats..." Robertson

"In Extension mine two men..." Dawes

"If you were injured..." Ibid.

"Before we got organized..." Robertson

"During the Depression we couldn't..." Ostle

"They let the single men go first..." Gilmour, 48

"When we used to be lookin'..." Edwards

"Those were tough times..." Crawshaw and Albert Tickle

"Miners received coal at a very low..." Johnston, 65

"We got a special permit..." Edwards

"If you couldn't find a job..." Lewis, 72

"We didn't make no money..." Arman

"Two of us had third-class..." Unsworth

"My wife used to worry..." Ibid.

"When the dust hit Saskatchewan..." Tom McDonald

"After they finished talking..." Dusty Greenwell

"In a way we kind of organized..." Gilmour, 48

"The boss would know..." Ibid.

"We were deliberately going slow..." Ibid.

"Dad was the first secretary underground..." Dusty Greenwell

"There were so many in a group..." Atkinson

"People wanted the union..." Sam Cameron

"We'd take a walk along the track..." Gus Cormons, CTHP 4051:27

"Each group had a leader..." Atkinson

"There was ten in our group..." Sam Cameron

"At first we had little meetings..." Horbury

"When it was undercover..." Mike Plecas

"These Yugoslavs were great ..." Atkinson and Gilmour conversation with author

"They were working for about..." Dawes

"When the Slavs figured out..." Atkinson and Gilmour conversation with author

"We were pushin' it a bit..." Robertson

"They used to publish..." Atkinson

"They brought all these guys..." Robertson

"One things they did have..." Horbury

"I can remember a meeting in the Finn..." Atkinson

"And so we used to distribute..." Ibid.

"Jack Atkinson was the leader..." Dusty Greenwell

"Bill Atkinson and Tony..." Mike Plecas

"The UMWA came in with..." Atkinson

"A UMWA organizer from District 18..." Gilmour, 48

"We come and we told Mum..." Mike Plecas

"We all joined up, every..." Ibid.

"Then came the day we thought..." Cormons

"You were supposed to pin..." Alex Raines

"We come out in the open..." Gilmour, 48

"They didn't go after wages..." Alex Raines

"We figured we needed the international..." Gilmour, 48

"See, I guess what they figured..." Mike Plecas

"Before we had the union ..." William Raines

"When they first opened it up..." Mike Plecas and Steve Krall, CTHP 4051:69

"I was working on haulage..." Ostle

"The morning shift had drilled..." Bryce, Mike Plecas, Brodrick, Malcolm Odgers CTHP 4051: 95

"I happened to look along..." Ibid.

"Boy oh boy... Ibid.

"In the meantime, all the horses..." Ibid.

"For once the horses and mules..." Ibid.

"Then they started taking up the men..." Ibid.

"It sure didn't take long..." Ibid.

"Finally it was our turn..." Ibid.

"Nobody was lost that time..." Ibid.

"The water was pretty near..." Armanasco

"And that uncle of mine..." Ibid.

"He was a tough old guy..." Dave McDonald

"After that we called him..." Atkinson conversation with author

"You could always tell who..." Atkinson and Gilmour conversation with author

"When I worked in Nanaimo..." Nelson Dean on Robertson

"All of a sudden there was money..." Carruthers, 23

"I was called up three times..." Thomas

"I was working in a shipyard..." Peffers

"People that had served in the armed..." Steve Plecas, CTHP 4051:102

"Well on Saturday you get off..." Gilmour, 49

"On Sunday morning Jim Gava..." Ibid.

"One of them must have stayed..." Loudon, 77

"The other fire boss was examining..." Carruthers, 22

"We didn't have six men for the rescue..." Gilmour, 49

"We had plenty of oxygen..." Ibid.

"I used to worry my head..." Mike Plecas

EPILOGUE

"So I hope that's going to interest..." Mike Plecas

"I worked thirty-three..." Robertson

"We had our good times..." Mitchell

"What time did I like best..." Galloway

Sources Consulted

Abbreviations:

Nanaimo Community Archives: NCA

Articles

Andrews, Margaret W. "Epidemic and Public Health: Influenza in Vancouver, 1918-19" *BC Studies 34* (Summer 1977)

Bowen, Lynne. "Goodwin, Albert." *The Canadian Encyclopedia*, Second Edition. Edmonton: Hurtig Publishers, 1988

Gidora, Mike. "The day that Vancouver stood still," *Pacific Tribune*, October 15, 1976

Johnstone, William. "Stalwarts of Darkness," *The Islander*, January 6, 1980

Norris, John. "The Vancouver Island Coal Miners, 1912-1914: A Study in an Organizational Strike," *BC Studies*, no. 45, Spring 1980

Roy, Reginald H. "The Seaforths and the Strikers: Nanaimo, August 1913," *BC Studies*, no. 43, Autumn 1979

Wilson, J. Donald. "Culture and Politics in Finnish Canada – Finns in British Columbia Before the First World War." *Polyphony*, vol. 3, no. 2 (Fall 1981)

Books

Barman, Jean. *The West Beyond the West.* Toronto: University of Toronto Press, 1991

Bowen, Lynne. *Boss Whistle. The Coal Miners of Vancouver Island Remember.* Lantzville: Oolichan Books, 1982

———. *Robert Dunsmuir, Laird of the Mines.* Montreal: XYZ Publishing, 1999

———. *Three Dollar Dreams.* Lantzville: Oolichan Books, 1987

The Canadian Encyclopedia, Second Edition. Edmonton: Hurtig Publishers, 1988

Encyclopedia of British Columbia, Daniel Francis, ed. Madeira Park: Harbour Publishing Ltd., 2000

Glover-Geidt, Janette. *The Friendly Port, A History of Union Bay 1880-1960.* Douglas R. Geidt, 1990

Gray, James H. *Booze.* Toronto: Macmillan of Canada, 1972

Griffen, Harold. *British Columbia. The People's Early Story.* Vancouver: Tribune Publishing Company Limited, 1958

Hagen, Judith A. *Chinatown, Cumberland*, Vancouver Island Regional Library

Harewood. Land of Wakesiah. Nanaimo: Harewood Centennial Committee, 1967

Hughes, Ben. *History of the Comox Valley 1862-1945.* Nanaimo: Evergreen Press (V.I.) Limited

International Correspondence School Reference Library. New York: Burr Printing House, 1905

Jamieson, Stuart. *Industrial Relations in Canada.* Toronto: Macmillan of Canada, 1957

Johnson, Patricia. *Nanaimo, A Short History.* North Vancouver: Trendex Publications, 1974

Johnson-Cull, Viola, compiler. *Chronicle of Ladysmith and District.* Victoria: Ladysmith New Horizons Historical Society, 1980

Johnstone, Bill. *Coal Dust in My Blood.* British Columbia Provincial Museum Heritage Record No. 9. Victoria: British Columbia Provincial Museum, 1980

Lipton, Charles. *The Trade Union Movement of Canada 1827-1959.* Toronto: NC Press, 1973

Mayse, Susan. *Ginger, The Life and Death of Albert Goodwin*. Madeira Park: Harbour Publishing, 1990

Morton, Desmond. *Working People*. Ottawa: Deneau and Greenberg Publishers Limited, 1980

Morton, James. *In a Sea of Sterile Mountains*. Vancouver: J.J. Douglas Limited, 1974

Norcross, E. Blanche, ed. *Nanaimo Restrospective. The First Century*. Victoria: Nanaimo Historical Society, 1979

Oliphant, John. *Brother Twelve: The Incredible Story of Canada's False Prophet*. Toronto: McClelland and Stewart, 1991

Ormsby, Margaret O. *British Columbia: A History*. Macmillan of Canada, 1958

Paterson, T.W. *Ghost Town Trails of Vancouver Island*. Langley: Stagecoach Publishing Company Limited, 1975

Reksten, Terry. *The Dunsmuir Saga*. Vancouver: Douglas & McIntyre, 1991

Robin, Martin. *Radical Politics and Canadian Labour*. Kingston: Industrial Relations Centre, Queen's University, 1968

Steeves, Dorothy G. *The Compassionate Rebel, Ernest Winch and the Growth of Socialism in Western Canada*. Vancouver: J.J. Douglas Ltd., 1960

Taylor, G.W. *Mining. The History of Mining in British Columbia*. Saanichton: Hancock House Publishers Limited, 1978

Thomas, Philip J. Thomas. *Songs of the Pacific Northwest*. Saanichton: Hancock House, 1979

Ward, W. Peter and Robert A.J. McDonald, eds. *British Columbia: Historical Readings*. Vancouver: Douglas and McIntyre Limited, 1981

Wejr, Patricia and Howie Smith. *Fighting for Labour. Four Decades of Work in British Columbia 1910-1950. Sound Heritage*, vol. VII, no. 4, Derek Reimer, ed. Victoria: Aural History Program Provincial Archives of British Columbia, 1978

Williams, David Ricardo. *Call in Pinkerton's*. Toronto: Dundurn Press, 1998

Pamphlets

Agreement Between the Vancouver-Nanaimo Coal Mining Company and the United Mine Workers of America, August 21, 1913. Private collection

Hedley, John. *The Labour Trouble in Nanaimo District: An Address Given Before the Brotherhood of Haliburton Street Methodist church*, 1913. NCA

Labour History, vol. 1, nos. 2, 3. A Provincial Specialist Association of the B.C. Teacher's Federation. Vancouver, 1977

Wage Agreement Between District 18 United Mine Workers of America and the Canadian Collieries (Dunsmuir) Limited and the Western Fuel Corporation of Canada, Limited. November 18, 1938

Government Records

British Columbia. *Annual Reports of the Minster of Mines*. Victoria: King's Printer

———. Attorney General. GR-0419 1857-1966

———. Attorney General Correspondence, Box 19 Pinkerton Reports on Strikers, 1913.

———. *Journals of the Legislative Assembly of the Province of British Columbia*, Session 1913, Vol. XLII; Session 1914, Vol. XLIII. Victoria: King's Printer

Canada. Parliament. *Commission Relating to Unrest and Discontent Among Miners and Mine-Owners in the Province of British Columbia*, Order-in-Council P.C. 2406, 1900

———. Parliament. Sessional Papers, vol. 13. Fourth Session of the Ninth Parliament of the Dominion of Canada, 1904. *Royal Commission to Investigate Industrial Dispute in the Province of British Columbia*. Order-in-Council P.C. 613. Ottawa: King's Printer, 1904

Letters

G. Arthur Bagshaw to Parker Williams, November 17, 1913, NCA

Thomas Isherwood to Emily Isherwood, August 15, 1913. Gladys Campbell collection

Israel I. Rubinowitz to Parker Williams, February 1, 1914, NCA

Henry Alexander Wilby to the author in 1981. Author's collection

Theses and Manuscripts

Bowen, Lynne. "Friendly Societies in Nanaimo. The British Tradition of Self-Help in a Canadian Coal Mining Community." Unpublished MA Extended Research paper, University of Victoria, 1980

Gallacher, Daniel T. "Men, Money and Machines. Studies Comparing Colliery Operations and Factors of Production in British Columbia's Coal Industry to 1881." PhD Dissertation, 1979

———. "Unhealed Wound: Changing Views on British Columbia's Greatest Coal Strike," 1981

Galloway, Marcia J. "Memoir" in Galloway private collection

Hagen, Judith Robins. "Cumberland Chinatown: The Origin and Demise of One Oriental Community in British Columbia." Unpublished paper, Simon Fraser University, 1980

Lockner, Bradley. "Miners in Nanaimo from 1880-1930." Unpublished essay, Simon Fraser University, 1980

Minute Book of the Enterprise Union No. 181, Western Federation of Miners, March 15 to August 9, 1903, Gilroy private collection

Smith, Brian Ray Douglas. "Some Aspects of the Social Development of Early Nanaimo." Unpublished BA Essay, UBC, 1956

"The Tipple," May 16, 1936 and June 13, 1936, Cumberland Museum

Wargo, John. "The Great Coal Strike: The Vancouver Island Coal Miners' Strike, 1912-1914." Unpublished BA Graduating Essay, University of British Columbia, 1961

"We Too," February 15, 1935, March 9, 1935, March 23, 1935 and September 21, 1935, Author's collection

Williams, David. "Memoirs of the Vancouver Island Coal Strike," B. Parker Williams private collection

Williams, Seriol. "Public Health Survey of Nanaimo." Student survey, Harvard Medical School, 1931

Newspapers

The Ladysmith Chronicle
Nanaimo Herald
Nanaimo Daily News
Nanaimo Free Press
Nanaimo News Bulletin
Vancouver Province
Vancouver Sun

Coal Tyee Society Interviews (CTHP: 4051)

Tapes and transcripts in Malaspina University-College Library and in BC Archives

Andy Adam
Ike Aitken
Steve Arman
Dominic Armanasco
Jack Atkinson
Victor Balatti
Rhoda Beck
Hjalmar Bergren
James Bevis
Herschel Biggs
John Biggs
Elmer Blackstaff
George Bodovinick
John Boloni
Wilf Brodrick
Ann Bryant
George Bryce
Margaret Cameron
Sam Cameron
Hugh Campbell-Brown
John Carruthers
Matthew Clue
Gus Cormons
Bill Cottle
Jock Craig
William Crawshaw
Harry Dawes
Alex Dean
Nelson Dean
Samuel Dean
Tom Dixon
George Edwards
"Wink" English
Harry Ellis
Dick Fiddick
Abbondio Franchesini
Elizabeth Inez Freeman
Jim Galloway
Scotty Gilchrist
Jock Gilmour
Lorenzo Giovando
John Gourlay
Archie Greenwell
Dusty Greenwell
Dolly Gregory
Henry Gueulette
Alan Hall
Clarence Hamilton
Ben Horbury
Mrs. John Hunt
John Irvine
Ernie Johnson
Tom Johnston
Laura Johnston
Eino Kotilla
Joe Krall
Mike Krall
Steve Krall
Ed Lee
Glyn Lewis
William Little
Robert Ross Loudon
Bill Lowther
Dick Mah
Dave McDonald
Tom McDonald
Effie McIntosh
Muriel McKay
Alex McLellan
Bill McLellan
Alex Menzies
John Milligan
George Mitchell
Charles Moore
George Mrus
J.L. Muir
Malcolm Odgers
Jack Ostle
Jack Ovington
John Pecnik
Johnny Peffers
Mike Plecas
Steve Plecas
Alex (Mike) Raines
William Raines
William Rice-Wyse
Johnny Robertson
Nettie Robertson
John Sandland
Mima Sheppard
Florence Simms
Arthur Simpson
John Slogar
Albert Steele
James Stockand
Dave Stupich
Thomas Terry
Lewis Thatcher
Marvin Thomas
Albert Tickle
Sydney Tickle
Waino Torkko
Jack Unsworth
Mrs. Frank Wall
Mrs. Evan Williams
Bob Winthrope
Albert Winkelman
Matthew Winkelman

INDEX

Author Photo: Dick Bowen

Lynne Bowen is the author of four other books of popular history: *Three Dollar Dreams; Muddling Through, The Remarkable Story of the Barr Colonists; Those Lake People, Stories of Cowichan Lake*; and *Robert Dunsmuir, Laird of the Mines.* Her books have won numerous awards including the Eaton's British Columbia Book Award, the Lieutenant Governor's Medal for Writing British Columbia History, the Hubert Evans Award for Non-Fiction and the Canadian Historical Association's Regional Certificate of Merit. She is the Maclean Hunter Co-Chair of Creative Non-Fiction Writing and the present Chair of the Creative Writing Program at the University of British Columbia. She lives in Nanaimo, B.C.